AN
OLD
ENGINEER
REMEMBERS

by

Meher   Kapadia

**A history of systems engineering and
software in very early
real time computer systems**

An Old Engineer Remembers

A history of Systems Engineering and Software in very early real-time computer control systems

Summary of the book.

This is an amusing true tale about the early days of computer control systems, based on the actual experiences and life story of an ordinary engineer. The book consists of technical descriptions of the systems of that era, intertwined around a memoir describing the engineering and business environment. It gives some idea of the fun and excitement involved in systems engineering, and will interest new and aspiring engineers. It may even have some reminiscences for older engineers. As a bonus this book includes descriptions of various international military,industrial and aeronautical computer based businesses together with their associated commercial practices. The individual chapters can be read separately as they are reasonably independent.

About the author

 Meher Kapadia has had a long and interesting career working for two large international systems companies, GEC Automation and CAE Electronics in the UK and Canada. His career also included a period working at SNC-Lavalin, a large diversified engineering and construction company, followed by some years as an independent engineering consultant. His appointments included positions in systems programming, engineering, marketing, executive management and consulting.

Retirement is his opportunity to do something else. Having retired, he now spends his time investigating and re-learning forgotten technical and scientific issues as well as keeping up with amateur interests in ancient world history . While spending time between two houses in the UK and Canada, he is still able to keep up recreational activities in cross country skiing, scratch golf, hiking and cycling. Very occasionally he even clears the cobwebs on his old dinghies while sailing on Lac St Louis in the west island of Montreal.

CHAPTERS.

1. INTRODUCTION

This is a short history about what early real-time computer control systems and electronic systems engineering were like. Those very early days coincided with a huge related and interesting   expansion of business and international systems work . Today computers and software can be found everywhere, controlling and managing all sorts of functions and devices. Even simple devices such as an oven will have an embedded computer, to say nothing about more complex machines such as the modern automobile which is crammed full of computers.   It did not start like this. At first computers were extremely expensive and could only be justified for the most important systems. One certainly could not scatter them around a system like confetti!

The initial idea for this story arose while reminiscing over my professional life as an electronic and computer systems engineer-programmer. I gradually realised that it had actually been a very interesting career. The more I reflected on it, the more I thought that it would be of interest to young people thinking about a career choice. Since then I have talked to many young students and fresh graduates, and in every case they seem to be mostly in the dark when it comes to career choices. The main problem is that they have no idea about what is the actual work they will be doing in their future career. I don't think university career staff either have a good detailed idea about what is the actual work done in industry. So at least as far as engineering is concerned, I have tried to describe in some detail what the work life of an actual engineer could be like.One of the unfortunate trends that often occurs, is that many young people who have taken an engineering degree, get a job in something else such as finance. There is a mistaken idea that actual engineering work is dull and badly paid. NOT TRUE !!! As a working engineer I had a ball. Even my average pay was not too bad. So read my story and decide for yourself.

I fell into this career almost entirely by accident. In the normal course of things, it is not something that would spring to mind, even among fledgling trainee engineers. In fact, the official divisions of the traditional university engineering faculties do not seem to have a permanent category for "Electronic Systems Engineering-Programming". This is not unusual, as many of today's new professions did not exist even ten years ago. At this point I should make it clear that there is a version of electronic systems engineering which mostly involves bureaucratic procedures and practices. In fact this is how much of the defence industry use the term. I use the term to describe the actual technical work, including both design and construction only, giving a lesser role to the organisational processes. The reason is because I cannot think of another composite term for all the general design engineering that is required when putting together diverse electronic and software components together. So when I use the term "engineering" I am mostly referring to the intellectual technical work, including systems architecture and design, not the associated administrative and support work. A good example of how systems engineering happens is given by the example of the cell phone. This was not a new direct invention, but involved using existing technologies to fulfil a perceived need. The innovation was foreseeing how telephones could be released from the straitjacket of fixed wires. One uses what has already been invented in new and ingenious ways.

As I started to write this history cum memoir, it became quite clear that unless I explained the technical issues clearly, this saga would lose a lot of its interest. So I have included explanations of how the technology was chosen and has evolved over these last 40 years. Towards the end of this book I have included some of my ideas about how this technology is likely to progress. Engineering is mainly a common-sense profession, and anyone with an interest can easily follow the technical discussions. I find that those who speak with jargon encrusted language and fancy words are usually the ones who do not understand the concepts clearly.

Engineering is a practice NOT a science. While it is useful to know some basic science and maths, the rest must come from experience. This is even more relevant for systems engineering, than some of the other fields such as electronic circuit design. Systems design especially is as much about intuition as about scientific and rigid rules. The main difference between scientists and engineers is not necessarily levels of complication or difficulty. Scientists can be working on problems that may turn out to be unsolvable, but that could still be considered a success. Engineers always finish successfully, though not always at the first attempt. Usually when engineers work on something new, there has to be a certain amount of trial and error, usually on actual live systems, while we find out what is going to be successful. So one should not be surprised if the final result is somewhat different from what was predicted, but the end result is a stable working system. One consequence of this is that innovative projects usually blow both the time schedule and the budget. This is quite normal.

These days, computers and electronic devices are ubiquitous. This book describes the period when much of this technology was invented. Of course the devices of this period were not as "good" as their equivalents today. By this I mean they were much more expensive, probably less reliable and certainly had much lower functionality. So bear with me, while I describe technology that will look hopelessly outdated. This is after all a history, and as such it is similar to other histories describing ancient technologies and practices! I hope when someone writes a comprehensive history of the "digital" age, they might find place for a short chapter for some of my type of experiences in real-time control. This real-time computing technology is still one of the most useful aspects of the use of computers.

However, almost all the modern design techniques and concepts are still mostly the same as those from that era. This is especially true for the design and architecture of complete systems. Some of the basic components are newer and have greater functionality. It is this greater functionality in the constituent components which gives greater power to recent systems. It also allows us to build systems which we could not dare build in those early days, though we did dream of building them. So I think this book will still be a reasonable history of systems engineering-programming in the early days of computer control systems.

Even though computers and software were invented during the war years, most of the computer related work was done in an off-line fashion until about the early nineteen-sixties. That is to say, software programs were written and run on a computer in a batch mode with results being printed out. This was fine for scientific, accounting and office type activities, as they could be done serially and could take plenty of time. However in the late fifties and early sixties, experiments were started in trying to use computers to directly control and monitor industrial,aerospace and military devices. To do this effectively, programs had to be run fast enough to complete in time so as to not disturb the physics of the phenomenon being monitored or controlled. As several software programs had to run at the same time, it was necessary to multiplex the programs all together in a single shared computer. This required new concepts in managing software within the computer with powerful and fast EXECUTIVE software that managed this continual allocation of software programs among computer hardware resources. The times involved were of the order of 1 second for most monitoring, and could be as low as 4 milliseconds for certain programs. So not only did the computers have to be fast, the software had to be efficient and not waste time or memory. The electronic interfaces between computer and real world machines also had to have the same requirements for speed, reliability and performance. This type of technology is generally called Real Time Computer Control. I was lucky to be among the early practitioners, so I have a good insight into how this technology has evolved. Even though systems work can be technically very involved, it usually requires only a relatively thin background application knowledge. To use a medical analogy, systems work is more like the work of a medical GP rather than a medical specialist. It is more useful to have a little knowledge about a lot of things than a very detailed knowledge about a small specialised area.

As I have spun this history around my own experiences, it is also something of a travelogue as I was sent to build systems all over the world. It gives some idea of the fun and great interest that such a career can have. An interesting point was that this period was also during the beginning of a new round of business globalisation. The airline industry had just started to introduce jet travel, and that made it possible, economic and convenient to do much more international work. Generally speaking international projects may have been fairly common in the civil and mechanical construction types of engineering, but was almost unknown for electronics and software. So in a sense my generation of control systems engineers were pioneers in designing and installing large electronic and software systems all over the world from bases in America and Europe. The developing world finally had the money and intention to buy only the best and latest technology. As I was fortunate in working for aggressive international businesses, I had my share of international exposure too.

The education of a systems engineer-programmer should be wide rather than deep. It ought to involve physics, mathematics and chemistry. These days some of the biological sciences would be appropriate too. Some sociology, and economics might help. As an engineer it is impossible to work in isolation, which is why one has to learn to get along with people and their idiosyncrasies. The fast pace in the creation of newer technical and scientific knowledge makes it very difficult to foresee what would be a good background education for a new engineer. When I started, software was not considered a particularly useful activity for an engineer. I doubt if I could have foreseen how much the field would come to dominate control systems and industrial simulation. In any case it is unlikely that my teachers would have known what to teach us about software at that time. The technology that will be used depends on what is available and ones ability to absorb new knowledge. In our era the base technologies turned out to be electronics and software. In pre-history our ancestors had to use smoke signals. A hundred and fifty years ago, railway engineers used mechanical signals or flags to coordinate systems. Now we use computers and CRTs. All engineers have to expect to keep upgrading their technical skills throughout their career, right up to retirement. There will be no let up. Otherwise one can expect to be dropped by the wayside.

Technology and society are very closely interlinked. Systems engineering is the discipline which takes into account the actual use that society will make of any particular technology. It also has to take into account the methods that people will develop when using a technology. In fact society does not necessarily use the best technology, but usually the most convenient one, (as decided by themselves). For a successful technology, the winner is decided by the users not the technical issues.

It is at this point that conventionally one gives thanks to everyone who helped in forming the book. However as I started counting, I could foresee that to be fair, it would have included a couple of hundred people at least, so I decided that would be absurd. I owe a deep debt of gratitude to all the teachers, colleagues, customers and mentors who saw me through my career. It is to them that I dedicate this book. Apart from the fun that I had, they all taught me a great deal, and especially helped me through my many inevitable mistakes. Almost everything I did as an engineer, was done as part of a team so really this history should be read as a description of joint group efforts. I also thank my family and friends for all the support that they cheerfully gave me. This support is crucial for a busy engineer, especially considering all the late nights and the unusual and strange places that I have dragged them to.

2.EARLY DAYS

This is how it all started. My family in India was typically middle class, but somewhat academically snooty. My father in fact was a Cambridge University alumnus, a Star wrangler in mathematics no less. As such none of the family had much respect for professions where one got one's hands somewhat dirty. My sisters and I were expected to keep up our family record for good marks in all academic subjects. Every thing else was secondary. So it was not too surprising that I managed to get reasonably good marks at each stage of school and college. However just to get some perspective, there was seldom even a simple screwdriver or hammer to be found in the house. Even the simplest leaking water faucet led to a big hullabaloo followed by emergency calls to various plumbers to come at once.

As a teenager I had ideas about joining the Indian navy as a career, but my dad was not too keen. I think he wanted me to do science at somewhere like Cambridge. However I managed to somehow sneak in the entrance exam for the military academy and nearly got selected. Unfortunately I had poor eyesight and in those days good cannon fodder material in India was plentiful, so I was rejected.

At Cambridge University I settled into studying engineering, as I had no idea about what else I could do and still earn a living. Cambridge had an excellent concept for what an engineer should learn. It was mostly about the basic sciences and very little of actual practice. In my early career this gave me a lack of confidence versus other graduates with more practical engineering degrees. However in hindsight it was the best thing that could happen. One never really has the opportunity to learn the basic sciences properly at work, but it is only at work that one can learn the most advanced actual practices of engineering. In that sense the universities are at a disadvantage and rather behind the times compared to industry. This is not always understood by the general public, who tend to assume that the university is where the most advanced work takes place. Especially in electronics and software, industry is where the most advanced stuff usually happens.

As I had some facility with mathematics, for my Masters degree I took Electrical Sciences (mostly electronics) as my major, so that I could get better marks. Usually I was pretty hopeless in lab work, but somehow the theory papers kept my head above water. I was still clueless about what an electronic systems engineer really did. I had no idea about the range of possibilities within industrial electronics or the defense industries. All I had heard about were stereo audio systems and other such consumer items. This didn't seem very interesting at all. To get an idea about what real engineers do, I took the opportunity to go visit some engineering companies in different parts of England. One visit was to ICI Chemicals at their enormous industrial complex in Billingham. It was by far the largest factory I had ever seen. In fact it was the size of a small industrial city. It was the same feeling visiting the enormous automobile factory of Ford in Dagenham, though it made me wonder how reliable a car could actually be! The other plant I visited was the aircraft factory of Handley Page near London. The contrast was quite dramatic. Everything in Billingham was big, automated and full of futuristic monster machines. The aircraft factory was quiet and people seemed to be doing all the work by hand, apart from some automatic welding machines. I was no wiser after these visits about what to do.

In those days Cambridge engineering students were expected to spend part of the summer holiday in a real factory, getting experience in an industrial setting. In my first summer I went to work at Davy-Ashmore's steel-mill manufacturing company in Northern England. It really was an eye opener. I had never seen such large pieces of machinery. The work-pieces were so big, they remained in a fixed place and it was the turning lathe that moved in towards the work. The bolts that were used to secure these machines were so large that it required two men just to lift and turn the wrench.

I was given jobs in the workshops and taught to do the work of a fitter, which involved using a steel file to make perfect shapes by tediously filing away on blocks of metal. I was also taught how to use a lathe and make a screw vice. One day while making the screw, my mind began to wander and I missed the point when one has to move the fast moving cutting tool away from cutting the screw. So the tool crashed into the metal base. I thought I had worked all that time just to produce some more scrap. My supervisor, a kindly old soul, just laughed and showed me how to hand turn a few more turns and produce a good enough screw with the damaged part being hidden within the base block of the vice. Experience will usually overtake book knowledge. Another advantage of working in these surroundings was getting an understanding of the real world where real people lived and worked. Universities tend to be in lovely surroundings and spoil their students dreadfully. I saw something of the class structure of the British environment. In the factory we had 5 different classes of cafeterias. As a boiler-suited, supposedly filthy handed person I was entitled to eat only in the lowest cafeteria. The others were for white collar secretarial staff, the engineers, the management and finally the executives. While talking to some of those who were allowed to eat in the better cafeterias, I discovered that the food was mostly the same, but we had better prices in the lower cafeteria. So there is some justice after all ! For two of my other long holidays, I also managed to hitch hike all around Norway,Sweden and Scotland. For shorter holidays, together with friends I visited France and Spain. In those days Spain especially, was inexpensive and as students that was important. I must say we had lots of holidays in those days! At work, especially in North America one starts with just 2 weeks off a year.

The next summer I worked at the Ferranti company's industrial electronic laboratory in Scotland. The Scots were really nice to me as they considered me a fellow ex-colonial of the "nasty" English. I even learned to like haggis from the company cafeteria. The company made instrumentation for industrial processes. My laboratory made an electronic box that measured the movement of an automatic drilling machine. I mostly did testing, and measuring signals using high performance oscilloscopes. As you will discover if you read this book further, I made my normal ham-handed errors while making a chassis cover. It was the usual cutting off too much metal, and required some help from my lab supervisor to hide the error with some putty and paint. I began to feel that I was not cut out to be a real engineer working with my hands.

I happened to be lucky in belonging to a student cohort who would graduate during a period when jobs were plentiful for graduate engineers. Just a few years later jobs did dry up and those who followed me were just unlucky in their timing. Over the years I have come to the view that most of what happens in life is down to luck or chance, rather than any particular brilliance on my part. This is a good attitude to have as it makes one philosophical about what happens and one should just get on with the job rather than worrying.

Fortunately I had done some courses in advanced control theory and mathematics which fitted my nature perfectly. The problem then was to find a company that allowed one to work in this area. It just so happened that my graduation coincided with a revitalization of the British Electrical industry. The English Electric company had just created a controls division to work with modern techniques ie. mostly computers and electronics. To staff up this company, they were recruiting fresh graduates, offering to train them in the various parts of the company by rotating the newcomers through each department. For my last summer before starting work, I decided to travel as much as I could afford, as I assumed one is stuck in place when working. So I travelled (camping) for several months all over East Europe and the old USSR with three other students in a battered old car. It was quite an eye opener, but it turned out I was wrong about travel at work. In fact during my career I must have travelled significantly more widely than most engineers into all sorts of obscure places and best of all at company expense. No more camping rough in the wilds of Russia or Scandinavia.

I started work with English Electric as a graduate trainee at Kidsgrove in the pottery district of England. I was later told that the company had chosen to set themselves up in this area as labor was cheap and well disciplined because of the long history of industry in pottery making. It is a good example of how things happen in real life. I still do not see the connection between pottery labor and hi-tech electronics, but I suppose there must have been some tax advantage too!

It took me quite a few months to discover that I really liked working with software. Initially work was quite dull as I passed through stints in the manufacturing areas, the test department and the circuit design department. While my middle-aged mentors were friendly and kind, they were usually too busy to find interesting tasks for me to do. Then I had a lucky break during a short period working in the process software group.

In those days, real-time software was still new for use in control systems, so most of the experienced engineers had never actually worked with software. As a consequence I don't think they understood how difficult it could be. My boss just handed me the task to develop a software program to analyze the statistical likelihood of there being particular isotopes from readings taken by a high speed chemical scanner. This was a new idea from our customer ICI Pharmaceuticals (Imperial Chemical Industries). The pharmaceutical substance under review was put into the path of radiation and the output was projected onto an electronic detector and screen. Each isotope within the substance spread out to a different part of the output spectrum and each isotope had a different weighting relative to its part in the substance. These detector readings were read into the computer automatically and it was my program's job to calculate the proportions of each isotope within the substance and print out the results. This was a way to automate what used to be a manual job within the laboratory.

As all this was done in real time, the program had to keep up with the scanner and finish the calculations in time for the next input. It was a great introduction to real time programming. I had to learn how to handle interrupts, interface with electronic devices and their protocols, manage errors, stream data, use the printer driver (this is not so easy as it sounds, I still have printer problems right up to the present day), interface with an operating system and so on. The actual mathematical calculation within the application was the easy part, as any real time programmer will tell you!!

I had never even heard of Assembler languages or for that matter, Operating systems. The nearest equivalent work that I had ever done was to write a simple Fortran program at university. Even that had not been too successful . The computer that I had to use was a specially designed machine, the English Electric M2140. In those days it was presumed that real time software would not run on normal business computers, as they would not be fast enough and have enough facilities to multiplex the many programs that had to run simultaneously. This issue of specialist computers existed at least until about the nineteen eighties, when normal computers became fast enough to use in real time control systems. There was also the issue of price, these computers were generally less expensive, though faster. For a long time in our profession we used to brag about whichever computer we were using as opposed to our competitors. It all sounds ridiculous today when virtually everyone uses the same PC type computers.

Computers essentially consist of electronic switches, registers (accumulators) and memory. With these building blocks the switches and registers are combined to form instructions. Together with the memory it is possible to create software sequences of instructions called programs. These programs then run the actual calculation required. In real time programming, the complete program has to take inputs from the outside world and process them to create the required output, fast enough so that an external observer thinks that the result was instantaneous (for all practical purposes). Hence the phrase "real time". This time also includes the time required for all the support programs that are needed for physically reading in the inputs and then sending the output results.

Computing is best understood as a series of layers. At the lowest innermost layer there is the electronic hardware. The next layer is a major set of software programs that manage the whole computer and its ancillary equipment called the "Operating System". Then there are other layers of programs that actually do the pertinent work, often called "Applications Programs". All these programs can be directly written as exact machine instructions. However this is very difficult for humans to do. A machine language program is just a series of numbers and it is very difficult to understand the underlying logic by simply reading it. So a program written with character symbol instructions or acronyms is used and that is translated directly into machine instructions for running on the computer. This is called an Assembler language program. A further abstraction can make the translation from more intuitively understood phrases. This is called a High level language. At the conversion from high level to lower level language some inefficiency is introduced as the conversion is done automatically by a program called a compiler. In the early days computers were not fast enough to do real-time programs in high level languages, so it was almost all done in Assembler language, even machine language occasionally!!

The other part of a real time system is electronic hardware that is physically connected to field devices to read and write electronically at different voltages. These electronic boxes are connected to the computers via special electrical wired links using a well defined set of programming steps called "protocols" to accept and pass on signals. In some systems the wired links are replaced by wireless radio type links. It is worth remembering that computers run on "bits" ie. ones and zeroes only. As in any program or memory store there can be millions of bits, even missing a single bit can cause a drastic error. So from early on the designers of computers had to use statistical maths to correct for potential errors. This means that data is always coded together with error correcting codes which are continuously used to make sure that nothing is lost. Similar techniques are also used for data sent to and from other electronic boxes and other computers.

Everything within and outside computers is continuously checked and checked again to make sure data is not lost. The actual hardware electronics generally consists of complex electrical circuits connecting different types of electronic components. The basic components are transistors, resistors,capacitors and inductors. These can be arranged with connecting wires to produce composite devices that can behave as digital switches,accumulators or memories. From these basics one can develop even larger devices that can be composed into logic sequencers, adders,arithmetic units and signal transformation calculators that are the basics of hardware digital electronics. Process control and measurement electronics take external world signals that are analog in nature and convert them into digital electronic signals using different "transducer " techniques. More modern electronics now involve Integrated circuits (IC) that are very complex semiconductor chips that contain all the required components etched directly into very small chips. These are inserted into thin layered tracks on a supporting PCB (Printed Circuit Board), The tracks represent the wiring. Hence the very small footprint of modern devices and computers. Another important part of all electronic systems are the power supply devices that usually supply clean low voltage DC power converted from normal mains AC power. Earlier hardware electronic boxes (chassis) were generally quite complicated as they had to do very detailed logic or signal measurement and conversion all within hardware only. Nowadays hardware is generally conceptually simpler as most of the complication has been moved to software running in the very fast, but simpler electronic hardware. The underlying basic system trigger and coordinating element is the continuous clock signal which keeps the entire electronic system in sync and progressing sequentially. Modern clock signals run at very fast rates (Mhz or even GHz). So modern hardware has to be well protected against high frequency electro-magnetic interference.

Protocols are one of the more obscure items in real time software,but are also one of the most troublesome programs. Initially every company developed its own proprietary set of communications protocols for connecting different electronic systems together. One of the consequences was that only equipment from the same company could generally be connected together. If one purchased equipment from outside the company one had to also buy a detailed specification of the protocol that the equipment could understand. It took many years for the SCADA (Supervisory Control And Data Acquisition) industry to eventually agree to use standard protocols, and these standards were generally invented by the international telecommunications industry. One of the techniques in specifying communications protocols was to put different interconnection levels in different layers of software. Each layer decides where to decode and send various parts of the data packet when sending data over complicated networks. These OPEN standards are hence very complicated to implement and take up a lot of computer processor time. In those early days it would have been completely impractical to have implemented such systems as the computers of those days were just not powerful enough. So all protocols were optimised for short processing times. Some parts were even implemented only in hardware to keep connection times low. Some special protocol types are called device drivers especially for computer peripherals and are usually supplied directly by the computer manufacturers. However the specialised field electronics protocols were done by us in the factory as systems suppliers. In fact a major responsibility of a systems supplier is to write and implement all the missing software so that a complex network of computers and electronics can all interface properly.

In a real time system one technique to intersperse different programs was to use electronic interrupt signals to hold a program. Then while waiting for the external hardware devices to respond, another program did something else in the meantime so that computer processing time was not wasted. This interrupt handling capability was a key element for real time control systems. Typically the electronic items that had to be managed in this way were backup memory ie. disks and drums, inter-computer links, printers and consoles, scanning electronics ie. RTUs (Remote terminal Unit) and so on. To avoid wasting time, interrupt program handling was the job of the EXECUTIVE software which had to work very fast,using special computer registers to retain the location of where the current software had been interrupted so that the EXECUTIVE could return when the connection was ready.So real time software was quite complicated and required the knack of managing to intersperse different software programs together so that everything finished in time and there were no unusual blockages. In those days one of the only ways to debug systems was to somehow visualise several things going on at the same time and anticipating what that might lead to. There were debugger programs that allowed one to single step through a program and visualise the results of that step. However when the whole system is running, debugging became a very difficult art and one would have to analyse the whole area of memory called a memory dump. The debugger programs did allow one to go through a series of stopping locations to step through entire parts of the program.

As a recent graduate, I was used to taking up a book and learning everything in it. So that is what I did. I just picked up the various programming and systems manuals and suddenly discovered the fascinating world of how computers really worked down to how each machine instruction operated. This really captivated me in a way that a high level language just cannot.

Being somewhat lazy, though mathematically adept, every time I had a problem to solve, I had picked up the technique of going back to first principles, rather than building up a repertoire of more involved techniques. This way of working is perfect for Assembler language programming. It isn't really difficult to use if one concentrates. Another advantage that I still retained, was that coming from the concentrated studying required for a rigorous university degree, I was used to hours of focused work. So it didn't take me too long, (about a couple of days) to actually produce my first Assembler code. In the industrial world this was considered mind boggling. No one would believe it.

Let me explain. When one is a student, it is all about results. Everything has to be done to a very tight schedule before exams and the only person who suffers if anything goes wrong is yourself. As a consequence university students are at one of their best, fastest and most productive stages in life for technical problem solving. Once one gets used to the pace of normal industrial work, which depends on others and is in a generally more relaxed environment (tea breaks,gossip etc), things take much longer. However I knew no better, I worked at the only pace that I knew, and got the reputation of being some kind of super star. I might mention it has been all down hill ever since then.

The other benefit of working in software was that I instantly became the "respected" company expert, merely by being the only person who knew anything at all about my particular program. Truly " in the kingdom of the blind, is the one-eyed man a king" ! In any case though I did not know it at the time, it launched me onto my long and fascinating career in computer control systems.

Another feature of those days, was that we mostly worked in isolation, as if we were doing individual scientific research. So our documentation techniques were almost non-existent. Much of the knowledge was in our heads only. This meant that every time some one left the company, it took ages to get back to where they left off. From an individual's perspective it was very satisfying as one could do a complete job all by oneself. The projects were small enough that a single engineer could do virtually everything by himself. People also tended to get very possessive about their work, and wouldn't tell other people how things worked. It must have been very irritating for our bosses, who had to pay the costs and report progress while mostly being in the dark about what was going on.

Computers and software are commonplace today, and nobody is much impressed if you happen to work with them. To get a flavor from those times, one might imagine that we were being inducted into a highly elite priesthood which no mere outsider could penetrate. It certainly did no good for our modesty either. I remember visiting an old uncle of mine a few years later. He took me to visit a friend's horse ranch where they bred race-horses. The gentleman politely asked me what I did as a profession. I told him I was a "systems programmer". This obviously puzzled both of them. They finally decided that it must mean that I worked with radio programs or at a TV station, and they began asking me about various singers and rock stars. Needless to say there was complete misunderstanding for a time while we talked across each other!

My next assignment, still as a trainee engineer, was to write an off-line program to automate the drawing of a control cabinet assembly. This is quite difficult to do, as one has to incorporate all the rules and quality assurance instructions while doing the general assembly drawing program. This was not very successful. I had to abandon the effort, as my previous boss came to get me transferred back to his group. It seemed that he was short staffed and had just received a huge project and he remembered my fast foray with the isotopes.

I might mention that this issue of automated drafting and auto CAD (Computer Aided Drawing) has really only been successfully tackled about 20 years ago after the advent of reasonably powerful computer work-stations and graphics. To give some idea of how slow it was in those days, I will list the sequence I had to follow. I wrote the program code and then got it typed up on punch cards. This took about 3 or 4 days, depending on how cooperative the punch card typists were. Being a mere trainee, they quite naturally put my work at the back of the queue. These cards were then taken to the company "large" computer, at a site several miles away and I did not see the result of the compilation (converting the hand written program into machine instructions) for about a week. Of course when it came back full of compile or run-time errors, I had to go through the whole cycle again.

A good habit that I picked up was to do a rigorous "desk check" of the program logic by hand. This involved mentally mimicking the actions of a computer, instruction by instruction, through each step and each cycle and writing down all the intermediate results ready for the next step. It certainly was good training in writing correct programs and was cheap too, as our department was charged for every second when we actually used a real computer. The final result was the automated printed drawing of the control cabinet with all the sub-assemblies (boxes filled with electronics) properly placed in the drawing and all the associated parts lists properly arranged as a family tree. When I left the program still had errors and I doubt that it was ever used by the drafting department.

Another interesting session that I had as a trainee, was in the electronic circuit design hardware group. No one at university tells engineering students how equipment is formally specified. I did not realize how the various interfacing and test instructions, protocols and codes have to be written down and agreed between systems engineers and circuit designers. To use an analogy, it was as if we were inventing a very simple but complete language for the very different electronic boxes to use, in talking with each other.The company was specifying a full series of digital process control interface circuits for the first time and there were a lot of arguments and meetings. As an ignorant observer, I enjoyed the fireworks, but did not appreciate the content until many years later when I myself had to specify a family of control equipment. This group gave me the job of testing the analog multiplexer circuits in a realistic manner including creating Earth faults and other errors. So I spent my time dragging long cables all over the labs and setting traps to catch problems. Then I stepped through a series of tests to look for errors in the equipment. I was also asked to write the test specification for the production department to use when the design was finished. For a time I was also sent to the production department and set to testing the analog circuitry (basically an analog computer) that was going to be used to control parts of a steel rolling mill. When I found an error, I was expected to repair the circuit card by replacing the failed component. This involved removing the component using a soldering iron and taking care not to damage the thin tracks on the card. Generally I ended up causing even more damage, till my supervisor told me to give up trying to repair cards. I was obviously still jinxed when it came to working with my hands.

One interesting quirk that I noticed was that hardware engineers seemed to like pub-crawling and drinking much more than the software programming departments. So while I was in the circuit design department, every week we all got together and hired a bus to take us around some cheerful pubs in the Derbyshire hills for a session of drinking and singing. I remember our group leader was particularly good with the piano.

To give an idea of how one lived as a trainee engineer in those days, I shared a rented house with 4 other young men. It was good fun, as we began learning how to cook and generally look after ourselves. Central heating was rare in the England of those days. I can remember waking up in winter with a thin film of ice on my blanket condensed from my breath at night!

I managed to save enough to buy an old car. At the time it was a choice between a 3 year old Skoda from communist Czechoslovakia or a 6 year old car from England. So I chose the Skoda. A big mistake. I have never had a bigger lemon. The silencer (muffler) rarely lasted more than a few months. This was compounded by the great difficulty of finding parts and dealers. We used the car to get to work and travel on weekends. It had an annoying tendency to overheat on hills. I can remember once just breasting the top of the hill on my way to Buxton and suddenly finding my feet covered with hot water as the internal heater burst. As I was short of money I decided to fix it myself, using a soldering iron and masses of solder to plug the leak. Of course I didn't think about the fact that solder doesn't work well when it gets hot. So the fix could only work when the so called heater was not too hot. I had many similar episodes fixing the silencer with the glue like "gunk" that was then available in shops. It really doesn't work. I eventually had to get a new silencer system each time. One interesting item in that "communist" car was a starting handle. The last time I ever saw such a thing was my grandpa's old pre–war car in India. Still it did save me the cost of replacing the battery, as every time the battery ran down, I just used the starting handle. It caused much amusement in the company car park when it happened, especially among the senior engineers who reminisced about the "good old" days of motoring before the war.

Unfortunately after I had been at English Electric for about a year and a half, the outside world intruded on our stable life. The British Electrical industry was going through a whole series of mergers and acquisitions, resulting in administrative chaos in all the departments of our company. The eventual long-term result was that AEI, GEC, and the English Electric companies with their subsidiaries (Marconi, Elliott, BTH, Metro-Vick, etc)were all merged together and various subsidiaries were broken up and re-parceled out over different sites in England. All the industrial control systems subsidiaries were put into a single holding company "GEC-Elliott Automation Systems", at the time by far the largest in the world. As I was working in our process control systems group I was sent off to the new process control company in Leicester called "GEC-Elliott Process Automation Ltd". My friends who worked in the power automation and metal industries groups were sent off to Rugby. Not much remained at Kidsgrove. Even the computer manufacturing company was sold to ICL. So we lost our operating systems group and all the knowledge that we applications people needed to keep the computers running. It did have the advantage of forcing me to become knowledgeable in operating systems, just to keep my programs working.

Another quirk of this set of company mergers was the issue of which computer would be the new company standard. Each of the constituent companies used a different computer. Elliot had the Elliot 905, an upgraded fire control computer from a tank gun. It was too small for most applications. Marconi had the Myriad. It was used mostly by the navy and not too well known. AEI used a good computer, the Conpac 4060, but it was a foreign computer built under licence. English Electric had the M2140, which was designed especially as a process computer. Everyone tried to push their own choice. However eventually the company kept using them all until they had designed their own new computer range the GEC 2050 and 4080. That was however far in the future. Among computer people, there is always lots of friction about which system is the best. As my career progressed I had to use many other computers.

As I was a very junior engineer, with no equity in a house or family, the company move did not bother me at all. However many of my older colleagues were most upset, and many left to set out on their own, rather than move as ordered. Later in life I have seen many periods when this kind of instability occurs for financial or business reasons. The end result is always the same. Projects go late and over budget, staff leave, chaos reigns. Probably the business reasons don't pay-off either. This is especially true for industries such as project companies that are heavily dependent on specific individuals.

Again in my personal case, it worked out well, as I went from being the most junior person to being the most senior. Of course, I was the only one left!!

3. The Land of the Midnight Sun.

Kiruna is a modern well organized little town above the Arctic circle in Swedish Lappland. I spent almost 4 years over there. It was one of the most interesting assignments that I have ever had. This town is all about Iron ore mining. There are two hills in the town, Kirunavaara and Luossovaara. I think the "vaara " means hill in Finnish. The area is right next to the border with Finland and Finnish is almost as common a language over there as Swedish. Both hills are almost entirely made up of iron ore and that is why the town exists. The mining company is called LKAB (Luossovaara Kirunavaara Aktie Bolag, ie short for the Luosso mountain Kiruna mountain company).I think it has been in operation for probably a century or more.

Kirunavaara is the site of the current major mine. Luossovaara seems to have been mined out. It still serves a useful purpose as the town's skiing hill with a small slope, enough for training runs. The mining is all underground. The ore is dug up and sent to the surface, where it is transported by train to the town of Narvik on the nearby Norwegian coast for export around the world. Some ore is first converted into iron ore pellets at a huge pelletising plant at the surface.

LKAB is extremely advanced when it comes to using modern technology. They are really confident and forward looking in the use of the latest automation techniques. That is how I came to be there. Our company won the contract for a project to automate the complete 540 level of the mine. This is the horizontal level 540 meters below the original top of the hill. I assume there must have been a pointed top. It is quite flat like a plateau now with some mine buildings established over there.

The way this mine works is as follows. A series of tunnels are dug at a horizontal level deep within the mine in the granite part of the hill. From this level, side tunnels are dug towards the iron ore, which is then dug out using blasting and large bulldozer like machines. This ore then has to be picked up and transported at this level and dropped into storage bunkers at lower levels. As the ore is heavy and the distances from ore face to the bunkers are quite far (several kilometers), LKAB decided to use trains to carry the ore horizontally. These trains then dropped the ore into the storage bunkers at a lower level,where large crushing machines broke up the ore into smaller pieces. Finally the ore was taken by "skips" (large containers in an elevator shaft) up to the surface. The whole process was to be automated with a minimum of staff . The reason they use trains has to do with the heavy weight of iron ore and the requirement for mining out a level as fast as possible. They think it can only be done with trains. Nowadays many mines use large trucks, but I guess they knew their business.

For a young automation engineer this was almost a dream project. Strangely however, few of my colleagues were interested in working on this project (at first). Most of them preferred to join projects which involved working with the main British power utility on Electric power grid systems across England.

At this time English Electric had won some quite advanced projects from the CEGB (Central Electricity Generating Board). They were mostly data collection type systems with relatively little actual computer based control. One of my colleagues who had a

Phd, was working on one of the earliest attempts to theoretically calculate power network security on-line using state estimation. I will describe this technology later in the book. Suffice it to say it had rarely been tried before anywhere in the world. I don't know if these projects worked out well. As most of these projects occurred during the chaos of the various company mergers, I probably doubt that it was completely finished.

Another advanced project was for the Midland Electricity Board (MEB), a power distribution company in Birmingham. The reason I mention this project is that it was probably twenty years too early. It involved putting a huge map of the city of Birmingham on a CRT screen and then scanning around it using a trackball cursor device to locate a position and then including a work order for the power line work crew. It was meant to replace a paper based work management system. As outsiders we gaped in awe at this new technology. Again I think the hardware design was too early. It worked well enough most of the time, but was expensive and not too reliable. I think the MEB operators probably went back to their paper and pen systems. Still it gives an impression of the excitement and interest in our office. This type of distribution management system was resurrected by me in another company twenty years later when the technology was better and cheaper. "What comes up goes around several times". Or does one say there is nothing new under the sun.

There were other systems of note. A data and control system for managing the coal handling conveyors in an enormous new coal fired power station at Drax in Yorkshire. I notice that even after many changes in the British utility sector, this station is still operating. My flat mate happened to be the main programmer on this system and we made many "Yorkshire" jokes while cooking dinner at night in our shared house. The five of us in that house, were quite opinionated and enthusiastic talkers. We even took a couple of weeks off driving around Ireland and camping in that rain soaked country.

One of the things I most remember about the trip was being asked by a kindly farmer if we wanted to watch the TV while the Americans landed on the moon. After that, back at work we all felt that our projects were quite behind the times compared to the hi-tech nature of the space race. It was only later that I learnt that most of the computer stuff that we did was far in advance of the space computers. I suppose NASA did not want to take on risky new technologies at that time. This is something that I have noticed during most of my career. Generally speaking, advanced Industrial electronics and computer equipment is usually in use long before government and defense projects dare to use the same stuff. This is contrary to the impression that is given in the public media.

A case in point is the use of computer based engine room data loggers that we were then installing on some Polish cargo ships. It was nearly twenty years later that I put in the world's first computers and software for propulsion control on a military ship. I suppose the competitive industrial environment forced companies to keep innovating. We also built computer data loggers for the UK nuclear industry, which must have been the earliest use of computers in nuclear stations. Years later I came across this equipment again when working with another company on nuclear operator training simulators.

To get back to the Kiruna project, the contract required us to supply a temporary smaller system quickly, while the main system was delivered in a couple of years. The temporary system was to only control the iron ore crushing system while the rest of the transporting was done manually. This was possible as only a few sections of the mine were ready. The mine was a high productivity environment where they would not be prepared to wait for us to develop the full system, and so were prepared to do manual workarounds. Hence the panic in our department to find someone to start work on the temporary system. That was my first real full time job.

The temporary system was to be based on a small industrial computer the M2112 . Today such a computer would hardly be recognized as one. It had only 16 available instruction types and a maximum of 16 Kwords (16 bit) of memory. However it was really designed to work only with 4Kwords at a time, the extra memory was available only as a kludge fix. It really was primitive, going back in concept to the dawn of computing machines.There were only 16 possible instructions (using the first 4 bits of the 16 bit word). This included register instructions. We even had to accurately hand-calculate the execution time for each instruction to ensure that crucial events were not lost (even electro-mechanical calculators were only invented some years later). All computers of this era were essentially RISC (Reduced Instruction Set Computer) machines, as the complex instructions of CISC (Complex Instruction Set Computer) machines had not yet been invented. So if required, we had to do the equivalent of modern micro-coded larger instructions directly in our software. There was no cache either. This very basic computer was excellent for learning all the detailed essentials of computing and electronics.

To start loading the programs into the computer there was a panel with switches and lights. The way it worked, one had to key in the actual required instruction as a sequence of ones and zeroes into the panel. This was followed by a pulling a load switch that put that instruction into the next available memory slot. The first program to be keyed in this way was a program called a Loader program which then allowed the real program to be loaded directly from a paper tape reader. However Loader programs were too long and tedious to hand key into the computer, so we invented a pre-loader, an even simpler program which then loaded the Loader. In those days being lazy, we sometimes even invented pre-pre-loaders. It was also a way to show off to each other how clever we were in using the least number of computer instructions to get started. As the memory was in short supply, we could not afford to waste memory to permanently store the loaders. So it was usual to use that portion of memory as a buffer for temporary storage areas when the computer was running live, such as those for printer output files. As a consequence, once a system was loaded, if we had a problem that required a reload, we had to go through the whole cycle right from the beginning. To give you an idea of the tedium of a complete reload, it often took a couple of hours or so. The main program was held on spools of paper tape with holes punched to represent ones and zeros. The tapes were produced by paper tape punchers on another larger main frame factory computer. A typical problem in handling paper tape was that it had to run properly into the tape reader or it all got twisted and tangled up causing the whole sequence to crash. Some days we could spend the whole day doing no other work than just getting a new set of tapes loaded. So progress could be very slow. We had to be very careful at each stage in setting up and checking programs as the fixes took forever. When we were in the field at the customer's site, the bigger computer was not available. So we had to keep several copies of the program paper tapes in case one broke. Another skill that we had to learn was to be able to fix a broken tape using a patching tape made of plastic and putting the correct holes in the patch with a special hole punching gadget. As a consequence we had to learn machine language so that in an emergency we could hand code everything in ones and zeros just like a real machine. Finally when everything was loaded, one pressed a start button (which jumped to the first instruction in the program) and the program started running. We had to find out if it was working correctly by observing if the pattern of lights on the panel was flashing in an expected sequence. It was a bit like driving a car by looking only at the instrument panel.

Of course before we got to the loading stage we had to debug our programs on the larger factory computer using an emulator program that mimicked the M2112 instruction set. So you can see it was a tedious exercise, made more tedious when we had to make a major fix at site far from the larger debugging computer. Kiruna was a full day's travel from Leicester, so we had to find a way to do fixes and patches by hand coding patch tapes using machine language as we couldn't run Assembler converting programs. Inevitably we made lots of mistakes when we had to do this. The most badly coded, mistake ridden program that I ever wrote was entirely hand punched machine code on site,designed to drive a mimic board. This mimic had been introduced while I was half way through the site commissioning. I will never repeat that way of working. That level of hubris was typical for a young engineer then. It was fairly common to put in temporary fixes directly into memory as "patches" ie. quite literally patches! This involved keying in fixes to code that had already been loaded in memory by overwriting them with the required changes, one memory location at a time. If there was not enough space available by just overwriting, then one had to put in a jump to spare memory locations where the patch was loaded followed by a jump back into the normal sequence. Over time the number of patches grew quite large and there was a danger that errors would arise from the confused scribbles documenting the patches. So it was good practice to re-assemble the code back at the office main computer after cleaning up the sequence of instructions. However there were always some errors that were fixed only at the end of testing. So it was not unusual for the system to be accepted with a small number of patches, on the understanding it would eventually be re-assembled. Today this would be nonsense as compilation is quick and easy, almost faster than patching and much less untidy. Beside it isn't easy to patch a high level language,as it depends on re-writing actual memory locations.

There were two of us allocated to this temporary project, a senior engineer and I. He worked on the more sophisticated programs to actually manage the crushing and blending plants. My job was to do the more boring and supposedly easier diagnostic programs and electronic equipment scanning programs. The basic executive already existed, capable of handling the 4K word program space, which we assumed we could easily expand to the 16K word space. Needless to say, it eventually took me ages to get the extension to work. This I only did by pretending that whenever we were in more than the original 4K we had to change pages ie. we had to introduce the concept of four 4K pages. Each program had to keep track of which page it was in and which page the data was in. It was even possible to actually overwrite any location in a program sequence with an actual new instruction or jump location . This was a technique to change the logic on the fly. It was just another trick to save memory.

My boss, the senior engineer, had hardly any more actual programming experience than me, or else he wouldn't have given me the tougher job of handling the hardware scanning. In real time programming these are usually by far the most difficult programs to implement, and to give them to a novice was not wise. It is the way of this world that the more glamorous work is often the easiest too. So I learned how to interface with real hardware by trial and error. As an example of the stupid mistakes that I made initially, I did a relay diagnostic test program that selected all the 128 control relays in sequence to make sure they were selected properly. Each relay made a clicking sound one after the other and then the program went around the sequence again and again. It was all very satisfying to me hearing the constant and rhythmic clacking sound, until the hardware test engineer panicked that I would ruin all his relays by running them constantly and he shut the whole machine down. I obviously didn't have the sense to do only a single test.

To clarify this control system. There was the computer attached by a series of wires called a telemetry bus, that connected in virtual sequence to a set of electronic boxes that took in signals from transducers in the field and converted them from analog or digital signals into computer words. Some of the boxes (called chassis) also connected to output relays via digital output signals to send controls. The electronic scanning was done automatically by this hardware, so it was quite fast. The computer software then scanned the main scanner to get the results from the hardware scan. Many years later, with faster computers, most such scanning is done by software, but in those days it had to be done directly by the hardware. In addition there were other devices connected directly to the computer such as a paper tape reader and a teleprinter for messages. Electronic boxes talk to each other by sending specially designed sequences of calls. As mentioned previously these sequences are known as protocols. For a very long time most protocols were proprietary. It was only in the mid eighties that the industry had the sense to work with industry standard protocols. This then enabled systems to be developed where parts were made by different companies. This is called using OPEN standards, and is a good thing as it creates more business for everyone.

A push button panel was put into a control desk to allow an operator to instruct the machine from a fixed set of possible instructions. Each button initiated a separate set of software instructions. The signal started a special program (one for each button). So for instance there was a button to set data into a computer file, another to delete data, another to flag or accept an alarm, one to print a report and so on. In some ways a panel based system is less flexible but it is much faster and easier for an operator to use. In any case it was just an incremental development from the days when all input and output was through panel buttons that actually were directly linked to hardware electronics. There are still industries such as the nuclear industry that insist on panels rather than software for important control decisions. It is certain and safe, though to a modern mind it might look outdated. There is a fallacy in our new computer era that the way to advance is flexibility. Thus the new computer designs tend to offer flexibility in how a user sets up a man-machine interface. For example in a PC, one can allocate any required function to a set of anonymous function keys on a computer keyboard. However I have observed that most users are happy with a fixed set of functions that do not need to be changed at all. It is faster to learn how to use such a panel and more convenient. Years later while designing a system for the military I had to change my mindset, as they insisted on a set of fixed function push buttons as the result of their actual research on man-machine interfaces. Especially during emergencies, operators don't have the time to go through "flexible" sequences of keys! Sometimes we system designers should stick with the older methods. The shortage of memory was so great we had to keep our operator messages as short as possible. We divided the message into a variable part relating to the item in question and a standard part that told the operator what was wrong. This type of optimising of data continued for many years until memory was plentiful and one could just invent reams of messages without caring about size. That is why I understood very well the year 2000 computer problem when it arose, as in that earlier era we were always trying to save memory by any trick that was possible. Incidentally I think this system probably would have also had a year 2000 problem, though my memory is somewhat vague.

There were secure power supplies to feed the system. This consisted of what is called an inverter-battery system. The incoming AC power from the grid is connected through some electronics to a system that converts AC to DC and can be used to charge battery systems . This then feeds the computer and electronics. Everything ie. inverters, switches etc are duplicated, so that the equipment is reliable enough to ensure the computer network does not ever (or rarely) fail. In the process of going through such a system, the power is also "cleaned" up and is much more stable, and so acceptable for use by sophisticated electronics running at very low voltages. The main computer system was kept in a typical air conditioned computer room. There was a long field cable that went from the computer room to our remote field electronics that were far away in other parts of mine near the actual machinery to which they were connected through various transducers and relays.

All systems, even today, work in more or less the same way, just using faster electronics and more sophisticated communications links such as fiber optics, but the principles haven't changed. The quality of the system depended on the ability of the field electronics to perform in the harsh environment of a filthy and cold mine for years on end. Good industrial equipment rarely fails. Otherwise one might as well forget its advantages and go back to old fashioned manual arrangements. A perennial problem with field electronics is their susceptibility to disturbance from Earth faults and power surges from the mine equipment around them. Much of the heavy mechanical machinery that the control systems manage, is powered by huge high voltage (into the kilovolt range) power equipment. The computer control equipment on the other hand runs at around a few volts, with some circuits working in millivolts. So you can imagine the potential for disturbance or damage. I will return to this technical challenge later on when I write about working in the super high voltage environments of 735KV electrical substations. In those days this type of electronic technology was called telemetry technology. These days it often has other names attached, but it comes down to the same idea ie. measurement or control at a distance.

A short description of the mine machinery being controlled is in order now. Here is what happens. An electric train loaded with iron ore is driven into a dumping area. This dumping area had no overhead power, so the train had to be slowly dragged over the dumping area one wagon at a time by a contraption of levers that automatically squeezed two sets of multiple rotating tires on either side of the wagon and this pulled the wagon through. Then the tires released and grabbed the next wagon. While the wagon was being dragged through, a lever released the hinged bottom of the wagon floor and all the iron ore in the wagon poured into the ore bunker below. Ore levels in the bunker were measured using special sensors. From these bunkers the ore falls onto enormous crushing machines that pulverize the ore into finer pieces which are then sorted into other storage bunkers via large conveyor belt systems. Finally automatic skips take the crushed iron ore to the surface pelletizing factory. Our job at this first stage was to fully automate all this machinery from the entry into the dumping area until the skips took the ore out. There were transducers and power relays to stop and start the machines and measurement devices to tell the computer the level of ore in the storage bunkers. Most of the control software consisted of sequence controls to start and stop the different machines depending on what the mechanical situation was. As the customer had experts on how the various machines had to behave, they gave us the required sequence control algorithms and safety logic, which we then implemented in software.

Even though I was not a mechanical engineer, after many months of working with the mine machinery I got a very good, almost instinctive grounding in understanding the detailed performance and operations of the mine machines. So in the end I was treated like a machine expert and often asked for my technical opinion in how the control and operations should be done. This is one of the hidden advantages of doing applications programming ie.long exposure to an application gives one confidence and specific knowledge in how a machine will behave. All this is very interesting, and over the years I have become very familiar with many different technologies and operations, simply by working for a long time with those technologies and applications. It is an excellent way to extend ones general technical knowledge.

Safe sequence control software programs had to be tested for all the activities. To help develop and test these controls my boss and I rigged up a hardware test simulator with lights and switches wired up on a huge wooden board. This was then used to check out the software in our factory by mimicking the actions of the machines, especially the crucial timing of machine activities.. Once we had, what we thought of as a working system my boss was ready to take the whole system to the field for installation and commissioning. I then looked forward to starting to work on the main computer system back at the factory as part of a new and much larger team, but it was not to be. My senior engineer submitted his resignation and was due to leave the company. It was at the time of the large reorganizations that I have mentioned. So the chief engineer was in a panic and the only solution he could think of was to send me (the novice) on my own, to pick up the pieces and get the entire system working, even though I knew nothing about the application or the design. There was great hilarity in the office as one of my pals (to paraphrase the usual saying) said " some are born to trouble, some achieve trouble, but you have had trouble thrust upon you" !! During this period all the in house projects were losing experienced staff, and the big bosses themselves didn't know if they would be leaving or were being sacked. The very senior executives who create this sort of turmoil are rarely called to account for the chaos they cause.

So on a cold dark wintry night after a full day of travel by automobile and plane, I arrived at Kiruna in Swedish Lappland, north of the Arctic circle. It was my first of a lifetime of company paid business trips to obscure corners of the world. The leaving senior engineer's expectation was it would all be working shortly and that I only needed to stay a few weeks. Absolute nonsense, even this temporary system took me well over 6 months to get going. I shall try and explain how everything went wrong.

I was given living accommodation in a modern hut shared with 6 other contractor's workers. Each of us had a private room and there was a shared general sitting area. As it was winter, the piles of snow covered the hut almost to the roof and our car had to be plugged in when parked, or it wouldn't start in the morning. I think the temperatures were in the -30C range most of that time, and that far north the sun came out for only a short time around midday. From our company there were three of us, my two colleagues were responsible for installing and commissioning the hardware, and my job was initially to commission the software and then the system as a whole. My hardware colleagues finished their job quite efficiently and quickly. They expected me to do the same so that all three of us could leave. No such luck.

I soon discovered that almost all the applications software did not work when tried for real. It is worth going into some of the reasons, as it is a common issue that arises in writing real time software. The problem is that when something is checked one program at a time in an off-line manner the bugs are easy to find and fix. However when the programs have to share a single computer and essentially run at the same time non-stop while continuously getting and sending data to field devices, it is very difficult to foresee every possibility. So only by trial and error can one slowly check things out one by one. This is a very slow process. Even when the first few programs are running correctly adding a new one can create unforeseen situations that mean we have to go back to testing from the beginning. When I called up the bosses in England to complain about what I had been given and asked for help, they could not send anyone and asked me to stay on and take as long as was necessary to finish the job. My hardware colleagues also left, and I had to take charge of keeping the hardware working too. In addition, I had an angry customer who was indignant at the fact that even this temporary system was going to be months late. In retrospect it was the best training possible to becoming a real engineer, being left to myself to get the whole system working.

The pattern of my life on site was as follows. In the morning after cooking myself a quick breakfast in my room I took the car up to the top of the iron mountain (Kirunvaara) and parked it near the pelletizing plant. Then with the usual miner's helmet and lamp I took a fast elevator down to the 540 meter level where the control room was. I started commissioning and testing by including each program one by one and then going through all the series of tests that I could think off. When they all worked using the wooden test simulator, I started the same effort one by one using the actual mine equipment. To do this I had to warn the mine operations management and get their help in placing staff at the various positions around the mine, to check that the correct response was being made by the computer.

Some of the logic was still faulty. For instance, at one point the tires that pulled the train through the dumping area pulled a few wagons through and then retracted. It was only after several minutes that we realised something was up when we got a call from the irritated train driver that he was stuck on top of the huge dumping hole and could not get out. A more serious problem occurred when the conveyor belt program got its logic sequence wrong. The problem occurred when the falling iron ore was supposed to drop onto a movable conveyor for taking it almost a half a mile over to another part of the crushing area. The program did not move the conveyor correctly and a whole train load of ore fell onto the conveyor rail tracks instead and jammed up everything. This led to chaos in the mine. Eventually a whole work gang of miners with pick and shovels had to go and manually dig everything out. So over the next few months I got to be quite well known in the mine and was friendly with the miners, and even began to learn some Swedish and a few words of Finnish too. We used to all eat lunch together down in the mine underground cafeteria near the control room.

For those readers who have never been in an underground iron ore mine, I should mention that it was a huge place, even so far underground. There were miles of road way and train tracks in enormous tunnels. In fact most of the miners got to their work station by getting a bus at the surface and driving down several miles to a junction where they even had to change buses to go further . It was to all intents and purposes a mini-town and factory hundreds of meters below the surface. The temperature in the mine was cold but not as cold as on the surface. The most fascinating thing I noticed was when occasionally the power went out and it became pitch black, one could not even see one's hand if one put it right in front of one's nose. Over the months I became quite blase about working underground and even didn't mind it when the elevator got stuck.

At the end of the work day, I came out and drove to the hut, where I cooked a meal for dinner, sometimes sharing with the other contractors. We might then go out on the town for a drink or walk. On weekends all the railway track laying hut dwellers got together for a session of drinking hard liquor, which went on until everyone fell flat on their faces. As I was not used to much drinking, it fell to me, as the sole person standing, to drag those huge six foot plus Swedes into their beds.

Life in the north is good fun if you like the outdoors. I bought some cross country and alpine skis and used to try skiing at weekends. I managed to learn enough to make it my lifetime hobby, even now in my dotage. I never got fond of that other absurd hobby that is ice fishing on frozen lakes, but I did join in watching the major winter ice fishing competition out in the wilderness, sitting on reindeer skins, chatting and drinking vodka.

After the hardware engineers left, I had to try and fix faults in the electronics by trial and error. It was a good way to learn about hardware issues. When the problem got too difficult, I had to call the head office and talk to experts on the phone, who led me through the tests long distance. There were days when I could be on the phone for as much as an hour at a time. My bosses thought it was still more cost effective, than flying an expert to the site. Modern electronics is fairly easy to debug at the PCB(Printed Circuit Board) level, simply by exchanging PCB boards. However there is a limit to how many spare boards are kept on site, so we had to even debug components on the boards, by replacing components or IC(Integrated Circuit) chips . The way this is done is by using an extension board that is plugged into the defective board's socket in the chassis. Then the defective board is plugged into the extension board. After this, the system can be run and one can trace the various active signals on a oscilloscope by using the scope's probes. To fix any problems, required good soldering skills and the ability to test PCBs using oscilloscope probes to monitor signals while the PCB was plugged in on the extension boards. During this testing I had to write simple test programs that continuously cycled through a sequence of instructions so that I could observe the effects on the oscilloscope.

During the later days on site, the electronic equipment kept failing in strange ways. This puzzle was eventually solved the hard way. Industrial electronics have to run uninterrupted for months, even years at a time, so there has to be a test that can prove this. The only test that seems to work is stress testing the equipment significantly more than is normal or reasonable. Thus we had to do a heat run. What this means is that the equipment was covered in large plastic sheets and electric heaters were put under the sheets as well. Then the heaters were run at high temperatures to get the equipment to fail. This allowed one to find the most susceptible parts and replace them. Then one should cycle the system back to normal temperatures for awhile, before doing the heat run again, This kind of cyclical testing continued until the equipment ran without fault for at least several days, at which point it was assumed that all the weak links had been fixed. I think that because of the rush to deliver the equipment as early as possible, we had shipped the hardware without a full factory test. So in the end it did not save any time or cost. My hardware colleague had to return and go through the heat test all over again. To keep him well lubricated in the heat, we kept him fully plied with many bottles of beer under the plastic. This led to some embarrassment once, as it coincided with a visit to Kiruna by our managing director. He however was quite amused to be introduced to my hard drinking, bare shirted colleague under the plastic sheet! In the search for quality, there are few if any short cuts. Since then I have always insisted on a full and complete factory test including a long time trial for every system that I have been responsible for.

However I am getting ahead of myself. One of the key items of documentation when writing software is to explain (in clear language) the purpose and logic of each sequence of computer instructions. That way the work can be passed on from engineer to engineer. This may be commonsense, but software programmers are notoriously protective about the ownership of their work, even though the software actually belongs to the company and not to them personally. The usual consequence is that the documentation is very scanty and it is awful to be given the job of finishing up someone else's program. There was a further problem with my ex-colleagues' software. As each instruction had to be followed with an explanation, the paper tapes of the source code became enormous, mostly by filling up with source code explanations. So to save on paper tape, he kept his instruction comments separately on a handwritten sheet of paper rather than on the compiled instruction printout. He had expected to put the comments onto the compiled code only finally after the system was ready. This was a foolish idea as the hand written comments soon got out of step with the computer compiled code, and he left me with long sheets of software and no coherent explanation of what the code did. So it took me several months to re-do large chunks of code, especially as it was full of errors due to not having been tested during continuous runs. This type of silly problem is why software managers these days insist on good documentation and testing techniques. It also pays to try and keep the same person on the job until it is finished.

Eventually I had nearly finished, when another problem loomed. The software team working on the main system at the factory had found that the software development would be running very late and that the temporary system would have to run for at least two more years on its own. To sell this idea to the customer, the general manager himself came to Kiruna and there was a huge uproar at a tense meeting with the customer. In the end the costs were re-negotiated and an agreement was signed. To celebrate we had a party at the mine's clubhouse followed by a game of bowls and hot saunas to relax afterwords. The general manager was very kind and complementary to me for holding the fort alone for so long. He was a cheerful Scotsman, and showed me how to enjoy a good malt. When the temporary system was accepted as complete, the customers even gave me another party at the club, before I returned to Leicester. On my return, it was like a hero's homecoming, with all my various bosses thanking me and giving me a good salary raise. Of course in those days inflation was so high that raise's value got wiped out fairly quickly. My next job was to do all the crushing and blending plant software plus the train control optimization strategies for the main system.

An interesting effect that surprised me was that, after several months in the field I had got used to doing most of my work standing up and moving around. In the office at the factory we sat at a desk all day and rarely got up on our feet. It was agony for a few weeks as my body got used to sitting down once again. A similar effect in reverse happened once more when I was back in the field and had to stand all day. Actually it is one of the advantages of doing a mix of site work and office work that one's body gets some change. It is just that I keep forgetting this changeover effect when it happens.

The main system was designed to have a very high availability, as the mine could not tolerate much downtime. As a consequence we had to develop one of the first ever computer control systems that duplicated all the critical equipment and changed over automatically when an item failed. When there was a fault in any part of the equipment, the system had to automatically changeover to the standby equipment without losing any data or control action. Even today this is not as easy as it sounds. There were two complete computer systems (M2140), which exchanged data through shared drums. In those days the memory consisted of a main (hi speed) memory based on magnetic core technology and the large storage memory based on magnetic drums. Magnetic drums had to be used as any other technology was too slow. Each drum was a rotating cylinder with a magnetic reader for each track. There were tracks etched all the way down the cylinder. That way any track could be guaranteed access in half the cylinder rotation time. This was essential as programs were too large for multiplexing solely in the main core memory. We had to develop techniques to overlay programs from specific areas of drum onto main memory ready for access fast enough to function seamlessly as a drum-core system. There were strict rules for memory allocation, with only the executive program and certain fast train control tracking programs that could stay permanently in core memory. Everything else had to work from the drum in overlay areas, keeping up with the speed required for reacting to events in real time. A further advantage with drums was that each track could be protected from corruption by pushing a switch per track. This then made that section of the drum read-only. So after we loaded our program code we set the appropriate tracks to read-only, leaving the data areas to be read-write. Modern computers today should have this type of protection too, it would avoid the current hassle of viruses and worms!

Thus I became quite knowledgeable in managing drum-core software. It was one of the many techniques that I have learned at great effort, only for the entire technology to get obsolete. Still it is quite neat how we managed to seamlessly couple the two types of memory. Each drum had direct hardware access to either computer, so we arranged for the active computer to read from the first drum, but write to both drums, so that both had the same data. The data and programs on the drum were arranged so that data could be called into core memory in sequence just as it was required. This kept up the speed of access as otherwise if we had loaded data at random, on average it would have taken us half the rotation time to just start to get the data. The data map of the drum was a key document that was under the control of the senior programmer in charge of the executive software, and we ordinary mortals had to negotiate and fight for our locations on the drum.

As part of real time event management, the system executive program could handle up to 16 hardware interrupts. As this was not enough for multiplexing all the various programs, some of the slower programs were all allocated to a single interrupt level and then called in a cyclical fashion in turn, but still fast enough for all the responses to be in real time. Essentially most of the real-time stuff had to respond with a complete real-time control within about 1 to 3 seconds. This time included everything from event registration, interrupt handling, executive call, memory handling, running application program, running electronic device controls etc. All this while sharing the computer with other background programs and effects. The fast memory technology was based on magnetic cores, so that memory kept its content even through power interruptions.

The train control hardware consisted of two systems. There was the normal telemetry system where a telemetry cable went all around the mine tunnels hooking up through the electronic chassis' to track-circuits and signals. It also connected with data from a remote ore analysis laboratory. This laboratory station took a sample of ore from each train, which was then used to work out which ore bunker the train should drop its load in. This ore quality measurement was used to direct the train automatically to the appropriate track by the higher level automatic train control process software. When a train is sitting on top of the track circuit it completes an associated electrical signal which is used to indicate track occupancy. This signal is also used to make sure that the computer does not allow the train behind to get too close by setting the associated track-side signalling lights to the correct setting. This telemetry system also allowed the computer to automatically set or change the track points so that the following trains went down another track.

The other system was a microwave radio system from one of our subcontractors. The tunnels themselves acted like microwave wave guide conduits so that the signals did not dissipate. Each locomotive had a special radio receiver that could detect a coded signal that instructed the locomotive what the speed should be and when to start or stop. To keep it simple there were only 5 speeds possible from zero to full speed. This is how digital control tends to be written. Most of us are used to analog ways of thinking, so we expect a continuous range of speeds from zero onwards. If you think about it, in a digital world it is easier to select a fixed set of speeds only. Not much point in getting to the in-between speeds, especially with trains. I have noticed over the years as control systems became more digitised, every option is selected from a pre-set schedule of possibilities. This avoids unexpected faults that could arise when selecting from an infinite set of analog possibilities. Many years later, I worked in the same fashion when digitising control of other machinery such as gas-turbine marine engines. The automatic train control software had to send these speed signals depending on the traffic. In the first year of operation there were drivers in the locomotive cabs, but once the system was fully operational, the drivers were removed and the system worked automatically in driver-less mode.

Our complete software team was organised into several sub-teams. One for the computer operating system,device drivers and basic integration, one for the human machine interfaces, one for the crushing plant control and one for the train control. At peak I think we were around 15 programmers. Since the total time that we took to finish was around 5 years , one can see how complex and large the system was. Incidentally the original plan was to finish in around two or three years, so one can see that right from those early days, software development has maintained its track record of gross underestimation and schedule overruns. Underestimation is a permanent problem even in today's software era. I think there are two issues. One is that new stuff requires a trial and error process to find out what will work. The other is the perennial optimism of technical people. You will keep seeing this conundrum as I continue to describe other systems that I developed or observed being developed in the future. A major consequence of this delay was that the temporary system had to last far longer than planned, and as a consequence I had to keep going to site to upgrade that system to keep up with the mine developments. In the end the temporary system was so useful and popular with the mine staff, that it took an awful lot of persuasion to let us convert them back to the final main system.

At the time when the main computers had to be sent to Kiruna, it was mid winter. The electronic gear was sent by truck, driving for some days through the ice and snow up to the Arctic circle. On unloading the boxes in the control room the technicians were most surprised to see the computer cabinets leaking large amounts of water!! It turned out that the air in the boxes had frozen and when unloaded it began to thaw and the water poured out. We had to wait nearly a week for the electronic cabinets to dry out, but the equipment worked fine when it was fired up. In my various occasions working on field sites I have seen some other similarly weird events. A field engineer must be prepared for whatever is thrown at him, even using drying towels. However modern control electronics is quite hardy,as long as one doesn't mess around with electrical earth fault or shorting inducing events.

The main system was gradually commissioned on site. Each sub-team was given a time schedule fitted around a 24 hour clock. This was done to speed up the process and not disturb the actual mine operations during the day. Most of the full systematic testing was done on the night shift. For a while I was allocated most of the night shifts. As a consequence I had a long 24 hours every day. When you are on night shifts, it is difficult to have a reasonable social life as one is asleep when the rest of the world is awake. However I was determined to not miss any of the skiing trips that the whole team used to make up to Riksgransen, Abisko and Kebnekaise (highest mountain in Sweden). So what I would do was go straight from the night shift to the train station with my skis. Get on the train up to somewhere like Riksgransen, ski all day and then catch the afternoon train back to Kiruna and a long sleep. Sometimes I was so sleepy by the afternoon, I nearly slept on the train and came close to missing the time to get off at Kiruna station. I suppose I could have slept all the way to South Sweden. I also learned to take my cross country skis for climbs up a mountain called Loktatjokko, halfway along the train track to Narvik. At the top there was a hostel for spending the night, but I would just ski back to further down the railway line at another halt to catch the return train. I had to be quite quick as otherwise I would have missed the train and got stuck in the wild for another 24 hours. I also did a cross country trip down the river valley to the Fjell stuga (Hostel) at the foot of Kebnekaise. The wind that day was so strong down the valley, that I could not stop. Every time I stopped skiing , the wind blew me off my feet. When I reached the stuga for the night, I even had a short spell of snow blindness caused by the blowing snow drying out my eyes.

After a while, I became more and more of a fixture in Kiruna, as I had to look after both the temporary system and my programs in the full system. We never seemed to be able to schedule me for a single trip where I could do all my work at one go. Hence I made a lot of trips by air and got quite used to the aeroplane trip through London, Copenhagen and Stockholm. For variety, I once did the trip by train though Gothenburg. Another time I took my car and drove all through Sweden in the middle of winter. In Gothenburg I nearly forgot to get some snow tires, but luckily remembered in time. As a bonus I met up again with some friends I had made years ago when as a student I had hitch-hiked over Scandinavia. They were quite amazed that this time I could speak some Swedish with them. My car was a six year old Mercedes, which was very comfortable, but did cause me some trouble when returning through Finland. I eventually discovered that the engine had a broken piston ring, but did manage to make it back home to England, before the car collapsed. I don't know what I would have had to do if I had broken down on the deserted wintry highways of Sweden. My return trip though Finland was long and quiet and somewhat boring. The country seems to be just miles and miles of forest,until one gets to the southern Baltic coast which is beautiful. Another side trip that I made was to the extreme North to Alta,Tromsoe and Hammerfest . I even managed to drive up close to the border with the USSR (Russia) and peek at that forbidden country from a distance. Near Kiruna in Northern Finland there are some Lapp villages where we went to watch dog sleigh racing. There was even an ice hotel.

In the summers we hiked in the mountains. As we were in the Arctic, the ground was wet tundra,and one could only walk with Rubber Waterproof boots, as otherwise one could get stuck in wet thawing mush. The other nuisance were mosquitoes. There were millions of them. It made walking in the open very annoying. At night there were parties we held for our friends (both English and Swedish), and I got quite a reputation for cooking very hot curries. Field work on site brings all the team members quite close together. I would recommend a field assignment experience to every fledgling engineer. I had so many local friends, that I even learned to speak a pidgin Swedish and a few words of Finnish. This even allowed me to play a practical joke on my Swedish colleagues in the mine.

This is what happened. One of my tasks was to develop and write the higher level decision programs to automatically route trains to the correct train track and bunker depending on the train traffic, the quality and the grade of the iron ore. I chose a modern control theory technique "dynamic programming equation model" for the software program, using as parameters a whole slew of inputs such as ore quality and track loading etc. The track signals had to be set up so that the driverless trains went smoothly without stopping to the correct bunker area. Before this automatic train routing program was ready, the mine control operator had to manually set up the track signals and instruct the temporary drivers as they approached the unloading area. The control operators got very good at it and were top rate at moving very fast through the decisions. I think it gave them lots of excitement on the job. Unfortunately for me I was trying to test my automatic program, but the control operators loved doing the job manually and did not allow my automatic program to run. I just couldn't persuade one particular operator to just wait and observe what the automatic program selected before he jumped in. So in the end, I decided to play a trick on him. I put in an illegal bit of program code that would run whenever he tried the manual approach . This program put out a printer message to him in my rough Swedish, telling him to stop pushing the buttons and let the automatic program finish. As I was on the night shift, I did the changes and went to bed. Next day when I came to work there was chaos in the mine. It seems the control operator was so taken aback with my pirate messages, that he thought the computer was alive and refusing to listen to him. These days we have all heard about hacking, but I think it was quite unusual then. To cut a long story short, I was told off by the mine manager for the disruption, but secretly all the miners who were in the control room had a great laugh at that poor operator's expense.

Incidentally my program worked fine and I did not need to try any more hacking. In modern times it is quite usual for us to get comfortable with the idea of a computer having an almost human like quality to respond sensibly. In those days computer-human interfaces were too clumsy and mechanical to be much more than machines. However that illegal hack showed me for the first time something about the ability of computers to behave like intelligent autonomous devices. These days modern artificial intelligence software is beginning to create computer systems that can think more autonomously. Later in life I did get some opportunities to try some of this, but that is for later in my story. One of the problems that we had then was the use of graphic CRTs to show the messages. In those days we only had the use of vector graphic CRTs and they were very unreliable compared to the yet to be invented raster graphics CRTs. As the screens got hotter, the characters drawn by the vector technique began to shake and shimmer all over the screen. On some days the operators gave up looking at the screens and used the fast golf-ball IBM printers instead!

Another thing that I learned was the importance of considering how an operator would be comfortable in using a computer system. A good real-time system is actually a close knit combination of man and machine. Each one brings their own advantages to the system. So merely making a system completely automatic is quite risky, as then the man is out of the decision loop. In an emergency he cannot understand what the computer did and why. So a good designer must keep the operator interested and in the loop. This is something that a young engineer is likely to ignore and get carried away with dreams of systems with no human intervention required at all. This is quite impractical as in the worst failure situations humans have to get involved in fixing the problems eventually. So the operator must be involved somehow.

After we finally finished the system we had a big party at the mine clubhouse and left the system to be run by the locals. I am told it ran very successfully for many years until the mine finished the 540 level and had to buy another system for the next level down. Unfortunately we lost the contract for that level at the last minute as our Swedish competitor managed to change the minds of the senior management of Kiruna. In those days it was more usual for nationalistic pressures to develop in the awarding of such key contracts as they were also a means of giving a local team experience in building large unusual systems. In fact on the strength of finishing Kiruna, we at GEC were one of the few international companies that could claim we could manage driverless trains through computer control. As a consequence when the city of Sao Paulo in Brazil went out for tender for a new control system for its underground metro system, GEC and Westinghouse were the only companies prequalified to be able to bid. Unfortunately we lost and I think that meant we never again did another train control system. On my return to the main factory I did some odds and ends on some steel mill automation software, while waiting for an opportunity on another big system. All of us from the Kiruna team kept up our earlier friendships and often got together for parties and pub crawls.

4. Some unfinished business with the CEGB

There is only one project that I voluntarily did not stay with to the end. Of course there are some others that I did not stay with to the end, but those were projects that were transferred to others by various bosses and circumstances beyond my control. The reason why I mention this is because I believe all good engineers should finish what they start, short of a disaster. It is a privilege to be given a system to design and not a trivial task. I realise there is a management tendency to change staff on a failing project, expecting that things will improve. I believe this is rarely an effective approach. Design and software details mostly exist in the designer's mind even on the best documented projects. So a more effective approach is to find a way to keep the same staff on, after maybe some small corrections. Usually designers are so attached to their work, a threat to take it away is enough to get the appropriate corrective action. So stick with it to the bitter end. If one is in management, look for the "finishers". They are worth their weight in gold. Avoid the people who start things but never stay on to finish.

The project that I refer to, was to automate the remote control and supervision of the Electrical high voltage transmission network in south east England. It was to be based on a control centre in the old town of St Albans. The existing system was an earlier design based on a truly enormous hard wired mimic board which used directly connected meters and switches to monitor and record the management of the power grid under supervision. All the coordination and control was done by telephoning each manned electrical substation The agency that ran the system was the CEGB (Central Electrical Generating Board)ie. the owner of the entire English power network. They had divided the transmission network into several regions, each with its own control room for remote supervision. Basically what the grid engineers had to do was follow the actual load growth and fall during the 24 hour day and make sure there was enough power generation available at the supplying power stations ready to pick up load as it came on. This was to avoid any fall in electrical frequency that would bring the grid system down. The transmission substations had to be monitored and switched according to where the power was to be delivered. All this is quite onerous and requires constant monitoring. In a failure or emergency the control room staff had to find a solution based on reconfiguring parts of the grid to bypass faults.

The engineers at CEGB headquarters were well advanced with ideas for using more modern techniques such as computers in grid control. I think they were probably the first to implement such a computer based management system at their national center in London, where they had implemented a fairly sophisticated system. So they were ready to specify a much larger system for telecontrol of the first transmission region which was based at St Albans. This earlier national system that they had built, had taken a great deal of effort, was years late and well over budget. So they thought they were being realistic in their requirements for the next system. Needless to say, these requirements get out of line very quickly and every such system seems to go way over budget and time. So it was with our system too. As I had only done most of the preliminary designs, I cannot say what happened in the end but even at the start, the project was grossly underestimated in both money and time.

This system was to be based on the GEC 4080 computer, newly designed at our sister computer company called GEC Computers. The parent company had finally decided to invent a new line of computers for the whole GEC company. This meant that each of the previous generation of computers from the original parts of GEC were banned from further use. So that was the end of the M2140. We had to learn a new assembler language, a new hexadecimal word structure (rather than the earlier octal structure), and a new set of hardware features. This computer had lots of excellent hardware features that are useful for real time process control, such as multiple protection layers, fast interrupts etc. Unfortunately the basic system software had been designed by a programming team who were more familiar with typical commercial batch machines. So the operating system and tools were too slow for use in a fast real time environment. As the principal designer, I came to the conclusion that I would have to somehow bypass most of the operating system and fiddle the basic system software to use the hardware functions directly. So naturally I made contact with the operating system team to find out details etc.

This created an uproar at GEC Computers. At first they flatly refused any help, until at a senior management meeting I made the case that otherwise the system would not meet the clients' requirements. So the very first 4080 system that GEC Computers made, did not use most of the expensively created general purpose operating system software. In those days computers had to be very fast to meet adequate real-time performance, or the system would fail. I would guess this bypassing of general purpose software has been fairly normal among us real-time types right up to the present day. The requirements keep getting faster and more stringent than the improvements in general purpose software. This implies that those who would do real-time system programming must get a strong appreciation of all the details of computer and electronic hardware. With that intimate knowledge, they can optimise issues of latency, speed and memory. I have noticed that each new generation of requirements usually more than doubles the sizes and speeds required.

There was also some idea to use a special purpose new higher level military programming language called "CORAL" as it appeared to be easier and safer to use for coding and better for documentation. In fact the GEC4080 Assembler equivalent language had some of the features of a high level language. At the time there was much talk about abandoning ASSEMBLER languages and moving to programs written entirely in higher level languages such as FORTRAN. I think it was too early and most such projects reverted to Assembler code as that was substantially more efficient. At the time I even believed I was probably also faster writing in Assembler rather than a higher level language, and told my boss that. Though he didn't believe that could be the case. It was only at least 5 years later that I worked on a project entirely written in FORTRAN. Usually on advanced systems one ends up making only incremental improvements between projects, rather than major changes. Even when we promise big steps forward we have to often go back to proven methods to meet performance requirements. One feature that we liked was that the basic executive program was actually coded in hardware. This made it much faster than a software executive, and had the added advantage that it could not be corrupted as the hardware was write-protected. Eventually on modern computers this type of coding does exist when instructions and even basic software is coded in various types of ROM (Read Only Memory) or cache memory. However in those days that was considered to be unique. It was also the first time that I came up with the concept of a software "process" and

message or semaphore passing as a safe method for transferring interrupts and events between processes. Much of this was actually built into the 4080 computer hardware and so was exactly what we were looking for to use in real time systems. I am surprised that modern computers do not use this type of design in hardware any more.
Computers such as the M2140 and the GEC 4080 had I/O (input/Output) processors in addition to the main processor. There could be several such I/O processors,whose task was to transfer data to/from peripherals autonomously from the main processor, which could then go back to doing other work. Basically these machines were multi-processor machines accessing shared memory in parallel and quite advanced, especially for process control type work. It was only in really primitive computers like my old Kiruna "friend" the M2112, where the main processor had to do all the work by itself.

An electrical power transmission region will potentially have several thousand analog and digital points to monitor across the various substations and power lines. In addition the system required many thousand more calculated points made up from algebraic equations based on the monitored points. This leads to a huge database that has to be updated in the order of every two to 5 seconds or so. In addition the system had to create a new software tool to set up all the static and parameter data for each point into the system. This "System Amendment" software tool was a new concept in those days. It had to be fast, user friendly and able to pick up any input data coding errors including logic errors. It was virtually an on-line compiler and required a great deal of effort to design.   These days such programs are often called database editors.

Another new feature was the use of alpha-numeric colour VDU terminals for the first time as control operator input devices. These VDU terminals had finally come down in price, and reliability and so could be used in a non-stop system. The VDU terminals were based on simple raster technology, so we could not do very fancy pictures, but with careful design it was possible to make simple power system one-line diagrams with just the alpha-numeric symbols. This was a huge improvement in man-machine interface (MMI)design flexibility, compared with the previous concept of hardwired push button panels. We kept the push button panels for some of the overview program calls and used the VDUs for all further details. This system was probably the first to use such a flexible MMI. We called the program that assembled the picture automatically a "Picture compiler". The main reason we had to try and automate the building of pictures was the very large number of possible line diagram pictures. So we assumed that if we had a compiler we could get clerical staff to do the actual work of building the picture by just inputting the initial data that was going to be used. The system was also supposed to interface directly with the mimic board so that all points on the board reflected the actual scanned status from the field. This automating of all the initial parameter data input followed by automatic software assembly of the entire structure of a system is now a normal and common way to develop SCADA type systems. In those days it was quite unique and might even have been the first time that it was done in such an automatic manner.

As one can see this system was a major change from the Kiruna type system, mostly in the enormous size of the data handling required. These transmission control systems at that time, did not use any sophisticated power system applications, but the sheer size of the system was so great that it was a big challenge to design. The system had the usual full redundancy as per Kiruna, and a large number of remote outstations in the field that scanned in data regularly over a mixed wire-line and wireless communication network. There were actually three control rooms, each with a number of control desks with VDUs and printers plus its own mimic board. There was a maintenance room with its own VDUs and printers to do the actual parameter and system amendment data input. In those pre-LAN days all the major computer network nodes were wired point to point across the system. The new GEC4080 had done away with our fast drum from Kiruna days. We had to make do with removable disks. These were much slower than a drum-core system, so we had to come up with new ways to optimise performance. The actual outstations did all the scanning by hardware so it was all hardwired and fast. These outstations were based on a new design process architecture called the MARCH series. I forget the significance of this acronym, but engineers are always cooking up strange names for their inventions. This CEGB system had a backup UPS (Uninterruptible Power Supply) power including the control center diesel generator, which would be used before the UPS batteries had run down.

The listing and printing of messages was also more advanced than Kiruna. One could for the first time get reports based on different criteria, including different types of alarm and event summaries. An important operator activity was registering and acknowledging alarms from the field. We invented new MMI approaches for this. Actually ,in the end, all we really did was copy the existing hardware approach to alarms and events. In those days each hardware point that could be in alarm, was directly wired to a set of lights called an annunciator panel. When it was in alarm the panel gave a sound alarm and the appropriate light began to flash until the operator went up to it and pressed it to acknowledge he had registered the alarm. So in the newer computer approach we created graphic type icons on the VDU screen that flashed when in alarm and changed colour. When acknowledged it went back to steady. The bigger problem was that we were changing from a parallel hardware approach to a sequential approach in the computer, so we had to come up with assumptions about what to do when scanned points kept changing too fast for the computer to register all the changes. In those days we did not have automatic filtering algorithms, and were still expected to not lose any events. In many power plants, even today, hardware annunciator panels are still in use.

All this was new then, but is quite standard in modern SCADA (Supervisory Control And Data Acquisition)systems. The problem that we were solving for the control room operators was really keeping track of and summarizing all the hundreds of activities and events. Otherwise the operators would have been overwhelmed with data. In that sense the system was a sophisticated business tool for managing power grid operations. In those days it was considered important to capture all the changing data without any filtering. More modern systems today use statistical filtering of information so as to avoid inundating an operator with raw data. Alarm and event fault analysis has become much more sophisticated now

Eventually the CEGB must have installed similar systems in each one of their regions. It was during this era, a major power outage took place over the whole North-Eastern American and Canadian interconnected electric power system. I remember reading a technical article on this and seeing the plans for system control that were being laid over there. Eventually all modern grids started specifying advanced control centres such as the St. Albans system, many with even more sophisticated programs for grid management. However St Albans was probably the very first, and hence a high profile system. At peak we must have had a team of over 10 programmers. The electronic outstations were of a new design invented by another full team of circuit designers. This terminology has been changed from outstation to RTU (Remote Terminal Unit) in North America and later that term became the common term around the world. The common terminology for such computer control systems became SCADA (Supervisory Control And Data Acquisition) systems, and some of the older terminology went into disuse.

A major effort in all SCADA systems is spent in developing automatic on-line software diagnostics for every piece of electronic or hardware equipment that is being used ie. the RTUs, the computers themselves, disks, memory, printers, VDUs etc. This is part of the hidden software that actually plays a major role in the highest quality systems. One of the consequences is that the programmer has to have a detailed knowledge of the internal design of the hardware electronic item. In fact the best diagnostic software program developers usually were the circuit designers themselves. In later years as a manager, I always insisted that only a trained electronic engineer would be allowed write diagnostic programs. Staff who only had some computer science or programming background could not be relied on to understand the subtlety of what the electronics was doing. So not only does a SCADA system have to continuously monitor the external plant under management, it has to simultaneously also monitor its own equipment too. Where hardware is duplicated we had to design equipment that would detect a failure and automatically changeover to the spare duplicate equipment. The most important such device was the Watchdog timer. This was an electronic switch that had to be pulsed from each computer at short intervals. If a computer failed to send its pulse it was assumed to be faulty and the switch immediately changed control over to the spare computer. It is important that a watchdog timer device is kept as simple as possible so as to avoid any condition arising that was not planned. Accurate time stamping is vital for a typical SCADA system especially in the power industry. Usually the computer time is synchronized from an external time source such as a satellite clock signal. This signal has to be the same one that is used by the high power grid network that is being monitored.

At about this time the very first microprocessor chips were being invented and of course they needed software to be written. So a turf war started within the company, as to which group was best placed to program the devices that had microprocessors. The hardware electronic engineers claimed that it was their job as at that time microprocessors were only used as sub-components in electronic boxes or PCBs to replace hardware logic. The software programmers claimed that real-time programming was a difficult professional activity and the engineers would not be skilled enough to do a good job. As is usual in large companies a lot of energy was wasted arguing these matters, and I think the problem was even taken up for debate by the highest level company executives. In the end practical considerations came to dominate. These days when microprocessors are embedded in something like PCB (Printed Circuit Board) logic, the programming is usually done by the circuit designer as that is considered part of the logic design. When a microprocessor is used for an application, the programming is done by the usual application software team.

As a result of their experiences on the previous CEGB system, the customers' engineers were very accommodating as we kept increasing our estimates for time and memory. In fact I remember that our displays hardware engineer worked for a long time to come up with the most carefully worked out lowest cost VDU type for their review. The customers didn't pay much attention and insisted on paying extra for more expensive "prettier" VDUs. As our project manager said in amazement "Money no object!!" I guess the CEGB was after all a state monopoly,where only the best would do.

At the time I did not expect to be particularly tied to power industry projects, but later in life I gradually seemed to be involved with mostly power industry projects. I had always assumed that process flow industries such as oil, chemicals etc used more sophisticated computer systems , but that is probably not so. The modern electric power industry is a much more sophisticated and confident user of computer technology than almost all other industries. As the electric power system works with the speed of light, the components need very fast electronics to monitor and observe events. Hence power network control requires the highest performing systems. The industry is usually an early adopter and does not hesitate to try advanced experiments. It is usually also much richer, so they can afford the best. I am glad I got involved early on in this industry.

As a consequence of winning this CEGB system our company was also selected by the Iranian Tehran Electric company (TREC) to develop a similar smaller system. So GEC created another similar team for that project understanding that this team would be using some of the software we were developing on the CEGB system. The consultants for the TREC system were a Montreal (Canada) company (Montreal Engineering) that had written the specifications and were supervising the design. This was the first time that I had come across this concept of a specifying consultant. In the past all our advanced systems had been specified by the actual customer, who knew exactly what they wanted. However less confident customers did tend to use consulting companies as an interface with specialist suppliers. I have always found it a nuisance to deal with a customer through an intermediary, as things can get lost in the interim. Sometimes it works quite well especially if the consultant is truly knowledgeable but that is not so common.

I have long had to deal with consultants in my career and will say more on this topic later. In those days I could not understand why clients specified systems in such great detail. I think it was a carry over from their earlier days when systems were all specified in hardware. In those hardware days the client used to be the expert in what the control had to do and what the available electrical signals could do. So it was natural that all they wanted from a hardware systems supplier were some bits and pieces of electronic equipment that they could then connect slowly as they wanted. Once we got into software the specification became much more complex, as the client had to specify right at the beginning the exact logic and sequence for all the control and management required. None of the clients understood much about software, and how software complicates systems. So inevitably the early system specifications were full of errors, and had to be changed during development. This meant that systems suppliers kept changing prices to reflect changed specifications. This led to a lot of bickering about who was at fault and who should swallow the cost overruns. In the end a negotiation led to usually splitting the difference, and software got a very bad reputation for overruns. So the clients thought that if they specified everything down to the very smallest detail there would be no more overruns. This was foolishly utopian and added a further problem, as different potential suppliers used different technologies to solve the same basic problem. It was unrealistic to specify so much detail before the competition was held. Personally I think it would have been easier if suppliers were kept on the hook for producing a system "fit for purpose" but letting them do it in their own way. Hiring so called "expert" consultants often just added an extra middle man to the discussions. Awarding a systems contract is a matter of trust, and one cannot do better by "nickle and dime-ing" everything. When one hires a plumber for the leaky faucet in the house, one does not tell him how to do his job (or at least one shouldn't) !

Another interesting issue that came up during the CEGB final bid process was that I was called in to give the software estimates in the absence of my boss who was away. This was my first tender review in front of the managing director (MD) and senior executives. I did not understand the political nature of such meetings and foolishly (though probably accurately) told those present that the estimates were quite wrong!! There was hushed silence as the engineering manager had to try and calm down the executives present. Eventually they made me describe in detail how I did my estimate and concluded I was more right than wrong and the MD refused to sign off on the price and insisted on a re-tender. That was also the last time they ever let me into such a distinguished forum. Eventually they must have agreed on a price because we did win the project. A major activity in systems design is to calculate all the main performance and cost items. As an example one must calculate the exact time taken to register an event at an outstation, followed by the time to transmit data to the master computer followed by the program run time. Similarly there are a host of other calculations that must be made for memory sizes, power requirements,number and use of interrupts,allocation of software between main and backup memory, allocation of functions between processes and so on. All this detailed work has to be done before actually starting the coding. In principle it is mostly commonsense, but these early design calculations and decisions are crucial.

One day my wife came across an advertisement for engineers and programmers for a Canadian company in the local British paper. She knew of my fondness for cold and snowy climates, so she suggested that I apply and I did, not expecting any answer. Much to my surprise I received an invite to an interview at the Canadian High Commission. In those days before professional looking CVs were made on computers, one just wrote up what one had done on an airmail letter in ordinary ink and pen. So my surprise was great that I got the interview. I had an interesting chat with my future boss and we mostly talked about what I had done in Kiruna. He offered me a job, and I accepted, not knowing what I would do or where I was going. All I could tell was that my salary seemed to virtually double (though maybe not in buying power of course). It was easy to make the choice, as at that time Britain was in a very bad financial state and inflation was very high. The new world seemed so much better to young eyes. Of course I realised that I was giving up one of the best jobs that I could think of, and who knew what kind of job was actually waiting for me. I guess I was lucky in hindsight, as things worked out much better than I could have expected. In fact if I had stayed on at GEC, I would probably have lost my job, as eventually GEC sold off or closed most of its constituent companies and is currently no more. I think the lesson from this event and others in my career is that one should generally grab challenging opportunities when they arise. They usually work out better than can be expected, as will be seen as my career progressed in Canada with a series of very challenging projects, each one bigger and tougher (though more interesting) than the previous one.

5. "O" Canada

On a cold and bleak November evening I landed at Montreal's Dorval airport to begin my long North American sojourn. It was only much later that I discovered that generally speaking November is the worst month for dreary weather in Canada. The beautiful and colourful fall is over. The equally beautiful and snowy white winter has not started. My new company CAE Electronics had sent the HR manager to pick me up and take me to the hotel where I had been checked in. It was a terrible introduction to one of the most picturesque cities in the world. CAE's factory was on a bleak industrial estate near the airport. So they put me up in a hotel just opposite CAE. It is strange how people don't think through simple things like where to house new people. They should have put all the new people that they hired somewhere downtown, as it would have been much pleasanter. Fortunately I had come on my own and my wife didn't join me until I was well settled some months later. As we engineers often have to travel for long periods of time to far off places it is a good idea not to disrupt one's family by moving them all over the place. The problem, as my wife reminded me, is that we have work to keep us occupied, but the stay at home family is trapped in a new place knowing no one. A lot of the wives of colleagues who came from abroad have confirmed to me how miserable they were when they first came.

At that time CAE was quite foolish in how they hired staff. For some reason they were reluctant to hire locally trained fresh graduate engineers, preferring to hire experienced engineers from Europe and Asia. As the main business was designing and building pilot training aircraft flight simulators, the only sources of trained simulator engineers were America, Britain or France, where the competing businesses were. American salaries were too high to compete against. French engineers would have had problems with the English language, so CAE was restricted to hiring from England. As England was in financial trouble then, it was easy to hire from there. One of the consequences was that for all practical purposes CAE was just like another British company, so I felt completely at home very quickly. I might mention that when I eventually did join management I made sure to finish with this foolish hiring process and usually only hired fresh local graduates. This worked out very well as it is not difficult to get fresh young engineering graduates up to speed. Their enthusiasm and energy more than makes up for their lack of experience.

Actually it was little more complicated than that. Canada was considered a good place to escape to from all over the world. So over the years, we at CAE hired from virtually every single country in the world. For instance just after the war, mainland Europe was in bad shape, so a lot of the early engineers were from Germany,Poland,France and Holland. Then the next batch were from England escaping high inflation and the low pound. Later there were groups from India and Pakistan, and so on. This wave of newer groups continued through my next 30 years at CAE, as different parts of the world came adrift eg. Lebanese, Egyptians, Africans, Chinese, East Europeans, South Americans etc. Eventually as the company became more famous, we even had lots of Americans. We used to joke that in every technical meeting there would barely be even one engineer who had been born in Canada. It was not only us working "stiffs" who came from all corners of the world. Our bosses were also just as varied. The president was Greek (born in India, so he often referred joking with me as "us Indians"). The next executive down was from New Zealand. The head of commercial simulator project management was Scottish, the head of manufacturing was English. The head of our military sensors division was Dutch. The chief engineer was Hungarian. The financial controller was Indian. The only Canadian born executives at the top level were the bosses of operations and finance.

This easygoing internationalism is one of the nicest things about working in Canada. It is also very useful, for a company like CAE is too large to exist by solely relying on the Canadian market. It had to become truly international right from its early days, so an internationally flavoured staff was a big plus. We could always find someone to speak and understand almost all the major world languages and cultures.

A few years before I had joined, the company had nearly gone bankrupt as new projects had dried up. It was to have a major influence on the thinking of the senior executive management. Years later I was talking with my old president, and complimenting him on his courage in taking so many risks with new advanced technical designs on almost every project. He laughed and told me that he thought it was the only chance that he would have to revive the company, and so he became comfortable with risk. As a consequence we engineers were lucky enough to do lots of interesting work all through our careers. It was also a great business strategy as we went from being the smallest simulator manufacturer to almost having a total monopoly of the civil simulator market world wide. It also made us the leading supplier of computer control systems for the world's largest and most complicated projects. I will say more about this type of strategy later on in the book, but I have come to think that if one does not take a reasonable level of risk, one will get nowhere quite soon. It is mostly the same for a personal career too. I have been well served by my tendency to take on losing projects that no one else wants to do. If one succeeds the kudos are even higher. Of course there can be failures too, but in the engineering world people have short memories and are generally optimistic by nature. Since then, I have recently seen that large unexpected changes can completely upturn normal businesses, eg. The Internet and international networking is upending older national only businesses based on a more stable model. So it is probably not as risky as is made out when an organisation moves boldly in an unknown direction. These major changes completely destroy the previous generation of successful companies, and the destruction occurs very quickly.

My first job at CAE was to help develop the software model and program for the de-aerator/heater system on the Pickering Nuclear station simulator. I had better explain. As CAE was the only Canadian company that knew how to build replica flight training simulators, it was chosen by Ontario Hydro (OH), the power utility of the neighbouring province, to produce it's first nuclear power plant operator training simulator. It was to be a joint effort with Ontario Hydro supplying the nuclear expertise and CAE the rest. This project was an experiment for both CAE and OH (Ontario Hydro). In Canada, OH was the main user of CANDU nuclear reactor based electrical power stations. As OH was an advanced user they had decided to speed up training of their plant operators by giving them a simulator for plant operator team training . They would observe how the simulation based training performed and then would decide if they wanted to buy other simulators.

At this time CAE were lucky to have bagged several different system contracts in addition to their usual flight simulator contracts. They were hiring like crazy, almost haphazardly. Many of my colleagues were new, mostly from the UK like me. I think before I go any further I had better say a few words about what a replica simulator is and how it is developed.

A replica simulator is a training device that "replicates" the machinery or in the case of a flight simulator, the full characteristics of an aircraft. Then it is used to train the operators or the pilots in a realistic device before they are allowed to actually use the real thing. It can also be used for training in emergencies such as loss of an engine in flight, or the loss of coolant in a power plant etc. Such emergency hands on training, of course, is not realistically possible on the actual equipment. So a simulator consists of the actual aircraft control hardware networked onto a computer system where detailed software equations fully model the various machines in the actual device. The equations are exact mathematical descriptions of the detailed physics and science that describes the machinery. As one can imagine each piece has to be modelled by a real expert else the device will not react reasonably like the real thing. Another advantage of simulation training is that it allows team training, which is not possible when training individual operators or pilots separately

In the past the simulation was done by a mix of electronic and mechanical hardware and some analog computers, but CAE had pioneered the use of digital computers to model the equations digitally in software. The method used to digitally replicate (analog descriptive) equations was to compute the equation in steps. By accumulating the steps one gets to a reasonably close behaviour of the actual equation. As long as the steps are calculated significantly faster than the actual physical device characteristics, the digital model will be a close enough replication of the physical behaviour. So one of the key requirements was to assess what the basic time step for the calculation was to be. The fastest time step was that used to characterise the flight equations at 20 or 30 msec (milliseconds) per cycle. For calculating slower processes such as boiler heat calculations a step of 50 or even 100 msec was used. In this era, floating point processors were far too slow when doing complex calculations, so virtually all calculations were done in fixed point format, and also in Assembler language.

Eventually many years later we allowed the use of floating point calculations when computers became faster. So in principle the entire simulation was divided into a set of software modules, each of which would describe one item of equipment in the simulator. For example, in a flight simulator there were modules for flight, flight instruments, air conditioning,engines, ancillaries, electrics, autopilot, radio, navigation aids, flight computers etc etc. In addition to the aircraft modules, there were simulation modules for the environment as well eg. APU, radio aids, weather, navigation aids, airport data, engine sounds and so on. Some items could only be simulated mostly by hardware such as the control loading devices used for simulating the effect of forces on pilot controls.The engineering organisation was also divided up the same way with a separate engineering group for each sub-system, hence being able to specialise in the sub-system. The underlying software executive just scheduled each subsystem program in a round robin fashion, at the appropriate time step. At each step the calculated value was stored in a shared data area called the X-ref. Hence that calculated value could be picked up by the next calculation as an input variable for its subsystem equation and so on. Each simulator was built with exact copies of the various aircraft parts eg. pilot controls, cockpit pilot seats, instruments and so on. We usually bought these parts from the aircraft manufacturer. The cockpit shell was built of a strong fiber glass surround based on a strong steel base. Sometimes we even sent engineers to crash sites to pick up crashed cockpits which we brought back to our factory in order to reuse some parts. When an aircraft part was too obsolete to buy, our innovative factory built an exact replica from drawings. Each major purchased part including items such as the computer, was specified in detail in a special document called a PSCD, used to negotiate the purchase and replace it exactly if it was ever needed again.

There are a couple of key subsystems that need to be explained some more. One such major subsystem is the motion system. Physically the motion system is a set of very large hydraulic jacks which hold up the entire aircraft cockpit. Using hydraulic forces it can move this cockpit with large extension movements at a fast rate. The human body reacts to movements in unusual ways.Early simulation engineers had discovered that if you want a body to feel a constant acceleration mimicking a fast accelerating aircraft, all you have to do is get the motion jack to move fast in the accelerating direction for a short spell, and then move the jacks backwards very slowly before doing another fast accelerating push on the jacks. The body inside the cockpit somehow only registers the forward jack motion and forgets the backward slow motion (called a washout). Using this we could simulate an aircraft takeoff very realistically while the motion system only moved the jacks by relatively short extensions. So from the outside a flight simulator looks quite weird, like a giant box on stilts moving up and down in jerky movements. We had the most advanced motion system called a 6 degree of freedom system ie. it could move the cockpit within 6 degrees of freedom. Some of the simpler simulators had 4 degree of freedom motion systems. Some simulators (usually military) were too fast moving to use motion systems, so the effect was simulated by using very powerful visual systems that fooled the body purely by manipulating what the eye saw. An allied system is the control loading system which is used to give a realistic feel to the movement of pilot control devices like the joy stick. The feel should stiffen or loosen depending on what the flight attitude and hence the flight surfaces eg. ailerons and rudder are set to. Modern motion systems are nowadays electrical rather than hydraulic.

Those hydraulic systems required careful handling as the pressures were quite high. Some years later there was huge emergency on the factory floor when a leak developed in a hose and the hydraulic fluid shot out at great speed and hit the roof of the "elephant house" just like a major oil rig gusher. It took quite a few days to clean up the mess. Incidentally as flight simulators had to have lots of vertical space to move in, the test site was a very high ceiling building, hence the name "elephant house". As our company grew, we had to keep expanding the test site areas to accommodate multiple simulators and other systems, all being integrated at the same time. So the CAE buildings looked quite strange from the outside. We just tucked the actual engineering office staff into all sorts of corners where the floor was level, as there was constant growth.

The electronic signals from the cockpit are led out through a "waterfall" of large cables that have enough slack to allow motion. These cables then attach to fixed computers and other interface electronics on the ground outside. To manage a training session there is a separate computer and display system called an instructor facility(IF). This facility allows a teacher to set up training sessions for the trainee pilot and follow his progress. The instructor can even take a smaller CRT facility inside the cockpit and manage the session from inside while watching the pilot. Early IF systems were purely panel based, but CAE had pioneered flexible CRT based computer IF systems. Of course inside the big cockpit box, it is fitted out to be an exact copy of the inside of an aircraft pilot's cockpit. CAE had a very forward thinking IF and computer systems group, so usually they were experimenting with the latest computer and graphics technology. One consequence of this was that CAE could also develop complex real time computer systems for other industries such as power control systems.

The other major subsystem was the visual system. This is a computer driven collimated projection graphics system that can show a very realistic view as seen by a pilot while flying. This was a very expensive system and we used to buy it from specialist companies and integrate it into our simulator. When I started work, the visual systems were only good enough for night time views of airport runways, to practice takeoffs and landings. By the time I retired, full economic daylight visuals had been invented by CAE and the training was even more realistic. It is so realistic that it can even fool the body into getting motion sickness. One consequence of our experience with this type of technology was that we were able to introduce very impressive graphic software into our eventual control system designs. Before we reached this level of visual technology, we had to make an enormous physical plaster model of the ground over which the training would take place and attach a camera to the simulator that followed the simulator flight. From inside the cockpit it was really remarkably realistic. One such system for a helicopter simulator sat for years on our factory test site as the Iranian air force who had bought it refused to pay for it during the turmoil in Iran after the shah. It always fascinated all the visitors to our factory.

Another visual technology that we invented for the Singapore air force was the dome visual. In this system the cockpit is put inside a huge dome onto which visual views are projected by very powerful projectors. The amount of data that would have to be sent was so great it would need impossibly powerful computers, so the dome visual works on another physiological trick. At any time the pilot's eyes are focused for detail only in a small area of his field of view. So only that field was shown in detail, and the rest of the view was shown without detail. To keep track of the pilot's viewing CAE had invented a head tracking system to follow head movements, and later even an eye tracking system for better detail. These dome visuals could even be used to train two pilots in one on one air combat, by projecting the correct computer generated views. In my early days military simulation was a small part of the company, but by my retirement it had grown to be equally as large as the commercial simulator business. Military simulation also included the simulation of various weapon,sonar and radar type monitoring displays inside the cockpit. For this we had specialised military simulation groups. Another consequence of working with the large quantities of data for radio aids data, airport visual data and instructor data was that at CAE we had developed many advanced real time database systems. This was to eventually also be a big advantage when moving to nuclear simulators and control systems. Database systems, including RDBMS (Relational Data Base Systems) have always been an important part of the CAE skill set. Another important skill we had was in designing very detailed on-line diagnostic and maintenance systems for all our equipment especially the electronics.

It all sounds fairly simple and in principle it was. However as the aircraft simulator had to exactly replicate the feel of the device to an experienced user eg. aircraft pilot, the equations had to perform quite exactly. There were all sorts of subtle effects that had to be catered for. So debugging the systems was very difficult. In addition the hardware that had to be designed, would have to be able to interface between the software driven electronics and the actual aircraft hardware such as the pilot's joystick. So the performance was characterised by both the software equation and the hardware interface together. A great deal of the final result was subjective and an experienced user would use his judgement to decide whether the simulation was good enough or not. The actual aircraft data that was used in the description had to be supplied by the aircraft or engine manufacturer. This data was used for the simulation equations. However there was no data for the really difficult situations when for example a plane would lose engine power. Intelligent guess work had to be used, based on just the theoretical physics.

Of course a nuclear power plant and a flight simulator are quite different, but there are enough similarities that many of the same experts could work on either. For example, the modelling of an aircraft gas turbine engine used the same physics as the modelling of the power station steam turbine, so the same engineer could be used to model either. Essentially those subsystems that were similar were done by the same engineering group. However we had to create new groups that had to be expert in new subsystems such as the nuclear reactor itself. This was where the user (Ontario Hydro itself) was expected to be the subject matter expert. As you can imagine many of the engineers in the various groups were highly qualified up to the Phd level, with years of expertise, enabling them to specify the models. Some of us had to rapidly go back to our university text books to understand the theory needed to do our modelling.

This led to my having to relearn my old thermodynamics from university days in order to do an effective model for the power station heat exchanger. Unfortunately by then I had thrown away all my old text books and so had to try and quickly find others in the company library. In the end it proved the point that even highly theoretical subjects from university courses can sometime be practical and useful. I am glad I had a theoretical and general engineering education, rather than just a practical education, as otherwise I would have had to go back to college to get the appropriate maths and physics background. I am all in favour of theoretical specialisation at university and leaving practise to the actual work environment. I would always recommend a young engineer to get as much theory as possible at university and not worry too much about the practical stuff. In any case most practical stuff has to be learned on the job at work, as each type of engineering business is different and the practical stuff cannot be found outside the work environment.

The computers that CAE then used for the simulators were TI980 machines, so I had to learn another Assembler language and another operating system. The software structure consisted of the TI Operating system executive being used to set up the system off line before starting. At start up, control of the computer was passed to a custom CAE written operating system called SIMTOS used to create the complete real time load module of all the run-time software which in turn ran under a real time executive called SIMEX. We kept 3 individual different dated copies of the final load modules (called grandfather, father and son). In that way as each one held an earlier version of the software, we could always go back to a previous version if the current version had bugs. This helped keep software configuration control quite stable. It was this SIMEX executive that controlled the flow of each model subsystem program. Everything running in real time had to be kept in core memory as there was no time to run as a disc-core system. At first I was quite scathing about the simplicity of the design compared to my background in disc-core design. However I learned quickly that in its own way the design was sophisticated and excellent for its purpose. To be able to adequately model a typical flight simulator one needed all the power of a single TI computer. However for the nuclear simulator, there were so many subsystems and data points, that we had to come up with the design of three closely coupled TI machines , with all three machines running in parallel. Otherwise the software steps would not complete in time to model the plant realistically.

This running of three machines in parallel was probably quite unique in that era. Each machine was initially loaded with its set of modules and shared data. As there was a minimal amount of data that all the machines had to share, this was kept up to date by inter machine transfer after every time step using inter computer data links. Basically one machine was the master and the other two were slaves, so the master had to wait until both slaves had completed their step cycle, data was transferred and the signal given to start the next step on all three machines. To debug one's programs they had set up a debugger that one could step through while observing the changes to the data in the Xref database. So one could observe how the equation variables changed or settled down. This was all visualized on a CRT screen. The whole debugging system was run on a larger computer where we ran a TI980 emulator. After we were satisfied with the mathematics and algorithms we could move to a test computer where we integrated each module one by one. A senior analyst /programmer was given the job of "integration specialist" and coordinated all the integration work. If a module did not work properly it was thrown out and returned to the designer to fix. This method was sensible and worked well considering that a large simulator might have a team of over 50 different analyst/programmers. The engineering group that looked after all the basic systems software had the job of keeping track of everyone else's modules and interfacing with the hardware.   For some time later on, I was the manager of this computer systems engineering department, before it was split off into a separate department just for control systems.

To give you an idea of how I fared, my first attempts to model the heat exchanger worked a little too enthusiastically and instead of cooling the steam from the turbines into slightly cooler but still hot water, appeared to cool the water so much it seemed to become ice!! Well one learns and eventually I got it right. As the client (OH) was a large bureaucratic organization, they insisted on very thorough documentation that followed exactly the same template for every bit of software. It was just the same level of documentation that had been used when building their own power plant. They would not compromise even one little bit in the quality of the document. So we had a QA (Quality Assurance) group which checked every line of code etc. This was probably my first introduction to real QA for software. In the past we did have rules and practices but were often careless about following rules, as long as the software worked and passed all the tests. At CAE this Software QA was also unusual as their usual clients (the airlines) didn't care how we worked, as long as the completed aircraft simulators worked well. It was during this period that software QA was coming into its own in various industries. The most hide bound were large organizations such as power utilities and militaries. At the same time as I was working on the Pickering simulator, the company was also designing and coding the first computer based ATC (Air Traffic Control) system for all of Canada. That team had even more stringent QA, which I can understand, considering the consequences if the system failed.

This ATC system was a series of distributed computer systems connected by a fast interlinking bus network. It was called JETS (Joint Enroute Terminal System). CAE had won the contract about a year before I joined. The JETS team was huge, at peak it may have had 100 software and perhaps 20 hardware engineers working full time for several years. It is an interesting system and was very successful in managing all of Canada's ATC control centers for many years without any failures at all. At a later stage in my career I was responsible for finishing the system, and I will describe it in another chapter.

In those early days, CAE was very busy, so my bosses kept giving me lots of different jobs on different projects. One such project involved the hand controllers and simulator for the "CANADARM". At that time the US space agency was designing a new concept vehicle to take astronauts into space, called the space shuttle. As space is a very hostile environment, with no gravity, they had to invent a technique to handle goods outside the shuttle in empty space. The Canadian government was interested in space research, so they decided to pay for and invent a special robotic arm which could be used to remotely move and place items in space. The prime contract for this device was given to another Canadian company SPAR, who in turn gave the contract for the actual controllers for the arm manipulator to CAE. In addition CAE was also given the contract for a simulator to train the potential astronauts in the use of this remote manipulator arm. A special electronics clean room was built at CAE to test and store all the relevant control electronics, with very stringent Quality control. The mathematical control algorithms were developed by professors from the local Mc'Gill university who came and worked with us as part of the team. In the end it was huge success for CAE and we got a great deal of publicity for it. It was very interesting going to the offices of SPAR in Toronto where we could observe the testing of the arm. As it had been designed to work in zero gravity, the arm had to lie on its side in the factory and only some linear tests could be done,as it was not possible to provide the zero gravity environment and 3 dimensional tests in the factory. So all one could do was very, very stringent quality control and hope for the best when the equipment eventually went into space. Some years later when it did, the arm was a great success and Canada became famous for space remote manipulators.

Another project that I started, but was transferred before I could finish, was to design the autopilot simulation software for the Tornado MRCA (Multi-Role Combat Aircraft)aircraft of the German air force. In an aircraft one of the key devices is an autopilot, that is used to automatically control the flight once it is in the air, without the pilot being required to do anything. A little earlier, I had also had to simulate the same type of device for a B737 flight simulator. In those days, the autopilots were essentially hardware analog computers that measured and then calculated the positions of each of the aircraft's flight control devices (ailerons, rudders,flaps) and then worked out the aircraft's attitude and flying characteristics and included measured engine speeds. Then depending on the settings that the pilot had input, the autopilot's job was to continuously maintain these control settings and only warn the pilot if something went wrong. So the simulation consisted of the actual autopilot hardware modified to interface with special simulation electronics that fed into the simulation computer. In the computer the autopilot simulation software equations were run to mimic flight and send readings to the aircraft simulation instruments. Essentially this was the same technique used for all the aircraft avionics and instruments. As the Tornado was a brand new design aircraft they had already started designing new autopilots that used on-board computers to do all the calculations digitally rather than using analog computing techniques. It was a hybrid(digital-analog) system, but many years later all modern aircraft have changed over completely to digital computer techniques for flight control and auto pilots. My background in digital computers confused me at first. This was because many of the aircraft instruments in the cockpit were called "xxx computer" as for example the "air data computer". So naturally I assumed that meant a digital computer not realising that most aircraft instruments were invented before digital computers were good enough, so essentially the word "computer" meant a device that did some calculations usually by some existing analog type feedback technology. So when other engineers were describing avionic instruments to me there were a lot of misunderstandings!

An interesting side issue was the requirement for all the engineers who worked on military projects to get an official security clearance, before one was allowed to view and work with the aircraft and weapon data. Those were the days of the cold war and Germany was a part of NATO. However in those early days, CAE was not very careful about security rules and it was only many years later that stringent rules and procedures were imposed. As I was new to Canada, it took quite a while to get my NATO secret classification, but because of the loose rules I was still able to access all the aircraft performance data so that I could design the simulation. These days I would have been chucked into jail for doing a lot less! CAE was a very easy going, practical place in that era. One could wander all over the factory without much problem. One of the advantages of being in flight simulators, was that when testing was off during the nights and evenings, we could sneak into the cockpits and learn to fly aircraft. The largest aircraft that I have flown was a Boeing 747,though I was not any good really. Often I would misjudge landing and crash the simulator. The test pilots who came up with us were good fun and very tolerant with us.

The first winter that I was in Canada was quite a hard one. I did not have a car and used to walk to work. Quite often the snow drifts were so great that it was actually better to walk than try to drive through the snow. One day around Christmas time I spent quite a long time helping to push stranded drivers stuck in the snow. Our company was quite hard nosed then, if you missed a day at work, no matter what the weather was, one did not get paid. So I went to work no matter what the weather was. One day it was so bad, only 3 of us on our project, turned up at work, as all the highways were blocked by snow.

As I was a temporary bachelor, without the rest of my family, I became very friendly with some of the younger engineers at CAE. In winter we used to regularly go downhill night skiing in the hills around Montreal after work. At weekends we would get together for all day cross country skiing along the trails and forests of the Laurentian hills. In spring and summer we used to camp, taking canoes far into the forests and come down the fast flowing rivers of the Quebec wilderness. This involved a great deal of "running" of the "white" waters. My canoe companion was an old hand at controlling canoes, so he guided the canoe and kept it straight, while I paddled furiously. We made a good team and rarely fell in. Some of our other companions were not as lucky and after a long trip we usually had several damaged canoes that we had to repair, as there was no way, other than the river, to get out of the forest. In the Canadian spring and early summer the forests are full of the most awful black flies and mosquitoes. In the forests one got completely covered. In fact it was so bad, I often preferred to chance my luck "shooting" difficult rapids rather than get covered with black flies while portaging the canoes through the forest. Years later I got to know the area around Montreal quite well and picked up a liking for mountain hiking in the nearby Adirondack, Green and White mountain ranges in the USA. It was only around 2 to3 hours drive to the nearest hills so one could make a day trip to the top of the high peaks at over 5000 feet.

6. A Gigantic country

Canada is the second largest country in the world. By air it takes well over 5 hours to go from the Atlantic ocean to the Pacific, and even then there are bits of Canada to the east and west. Over the next few years I managed to work on projects in virtually every province of Canada. Living in Canada is a bit like living in an ocean (but an ocean of land), the towns and cities are far apart from each other rather like little islands in that "ocean" of empty land .

One afternoon I was lined up to collect my lunch in the company cafeteria. As chance would have it, my boss , the chief engineer, happened to be in line just in front of me. He looked very worried, but his eyes lit up when he saw me. He had just suddenly remembered, that we had talked during my job interview, about the different SCADA (Supervisory Control And Data Acquisition) systems that I had worked on in my previous job in England. In hind sight that was a crucial moment in my career.

Here is a bit of background. CAE had tried to dabble in control and SCADA software systems as a diversification from just building flight simulators. However this business was not functioning well, as no one had the appropriate background. As a consequence the couple of projects that they had won, were in very bad shape, with enormous cost overruns, very late and yet not working. One of these systems was for the computer control of a brand new Oil fired power station near Saint John in the Canadian province of New Brunswick at Coleson Cove on the Atlantic shore. The system had actually been shipped, even though no one could get it to work and several of the engineers assigned to the project had resigned and left in disgust. Basically the system had been abandoned and the customer was threatening to sue CAE for damages. Our anxious company executives were bearing down on my boss to do something quickly. So that is how the trap shut on me that afternoon in the cafeteria as I casually agreed to do the job of finishing that project on site in New Brunswick. It did not seem a great inconvenience as I was new to Montreal and as yet did not have a house, so there would not be a big upheaval on the home front. In addition site wages were good, and that gave me a chance to save some money. The project itself was a real mess.

The first thing I had to do was to collect all the documentation that was available. The hardware and circuit card design drawings were in good shape. Unfortunately the software documents were in bad shape. I managed to get a complete listing of the entire software code that was in the system. Then I tried to understand the existing programs, starting from the homemade executive and the computer hardware. The computer was a Varian 620 computer. Here a diversion into the history of this design is in order.

Just a couple of years before I had arrived at CAE, they had won a project competition to design a new computer control system for controlling the Canadian nuclear industry's first CANDU reactor power stations. The plan for these stations, was for the client AECL (Atomic Energy of Canada Limited) to do the applications software and for a systems supplier (ie. CAE) to design and build the complete hardware and executive software. In addition it was required that we provide a special real time test and diagnostic program that would demonstrate that the complete system's hardware worked correctly over a long test period of several months. As this was the very first time in the world that a nuclear reactor was going to be actually controlled and monitored by computer software, the very highest reliability and availability standards had to be met.

The CAE design consisted of dual computer systems with a duplicate set of plant interface electronics. Each hardware interface system was independently connected to its associated computer. The fail safe nature of the design meant that each computer scanned all the hardware at the same time, and each field input (analog or digital) was read in parallel into each computer. This meant that if any element of one system failed, the other could take over instantaneously without losing any input data. This design also had the advantage of instantaneously keeping duplicate copies of all the system data without the complications of transferring data between computers. The control signals were also sent out in parallel, however there was a switched panel that only allowed the controls from the active computer system to physically reach the real control relay. This relay then controlled the device itself through a latching power relay. Basically this design was then sold to Coleson Cove as a thermal plant control system. However it had to be modified according to the process control requirements of an oil fired station. The basic software was taken from the nuclear DCC (Digital Control Computer) system design. The applications were all new and that was where most of the problems lay. There was also some modifications required to the analog input system, as the DCC worked with transducers that used 1 to 10 volts range. The Coleson cove system needed to work with RTD devices that worked over a range of only 50 millivolt. This caused many hardware problems for me in the field as this analog system worked very poorly. It took many visits by the circuit design staff before it was reliable enough.

Another issue was that the actual original nuclear DCC was not quite complete either. So in a sense both systems were development systems with still a great deal of required experimentation. Eventually the nuclear DCC business became a very successful business for CAE as the future systems were all copies of each other. These systems were installed at nuclear power stations at Pickering near Toronto, Bruce on the peninsula at Lake Huron in Ontario, Gentilly on the St Laurent river in Quebec, and Point Lepreau further down the same Atlantic shore as Coleson cove. In addition to these reactors in Canada these systems were also installed at reactors in Argentina, Korea, Italy,China and Romania. The above mentioned Coleson cove system was an orphan design and never repeated.

After my initial analysis I did an estimate of what it would take to finish the job. There were many software modules and each one would have taken a few days to re-test, modify and document. So it looked as if we would need a team of around 6 engineers to finish it in a few months. Unfortunately these plans fell apart as none of the other engineers allocated to the project wanted to live in Saint John. In the end I drove down on my own and set up the field office in the power station, near the station control room. We then planned that I would do all the design and modifications on site and then courier the changes to Montreal where the rest of the team would re-compile and test off line on one of the DCC systems that was still in the factory. Then the changes would be sent back to me for final test on site. If something was beyond my skill set, they would send specialists down for a few days to Saint John. In the end it took over a year to finally finish this system, and satisfy the customer.

Some of the peculiarities that I had to struggle with, were quite unusual. As the station was still only half finished, the construction crews and their dust began to coat the electronics and cause all sorts of breakdowns. At one point we even had to replace some of the wiring as a number of rats had managed to burrow into the cabinets and cake the insulation with their urine, which then kept creating electrical shorts. There was a remote third computer to monitor the electrical sub station on a hill about 1 kilometer away. The design was such that to restart this computer one had to climb the hill up to the substation and reset the computer, which made testing very irritating. This method of networking a remote computer was one of the early attempts to network real time computers over serial buses and longish distances. It eventually worked but was quite fragile much of the time, and very difficult to test. A further problem was that there was no disk memory in the switch yard computer, so the remote computer main memory had to be reloaded long distance over the serial link from the master computers in the station control area. Again this slowed down testing and debugging. I think in those days all computing hardware was considered significantly more expensive than engineering labour,which is why we spent so much in trying to save hardware costs.

There were no good tools to debug the system whenever it crashed. All I could do was take a complete dump of the entire core memory after the crash and then scan all the printed (octal) memory location contents to look for anomalies that might have triggered the system crash. In order to work out the possible crash scenarios, we had to mentally try and guess what had happened and in which sequence. It all depended on intuition and guess work. There was no way of systematically testing potentially damaging scenarios. A colleague of mine, who came to help me on site, was exceptionally talented in mentally debugging software. It was almost like magic, he would sit at his desk and just keep thinking about a problem for a while. Then he would get up tell me what the problem was and sure enough, he was usually right. It is episodes like this one, that have convinced me that real software talent is innate, though rare, and not something that one can learn. Its rather like being good at mathematics. In the future, as I was promoted to higher and higher management positions, I always made sure to nurture the few "naturals " that we had.

This system was the next step in what a computer control system looked like after my previous systems. There was a control desk per turbine unit (3 units). Each control desk had a raster CRT and printer with a control panel. There also were analog trend recorders that could be used to generate trends on any of the analog systems being monitored. The CRTs could show black and white symbol diagrams in addition to the usual reports. As the system was an administrative as well as a monitoring tool, we had to invent some special software to calculate the performance of the power station including the station heat rate. There was a scrubber unit so we also had to calculate the environmental parameters to keep the plant operating within its legal environmental limits. Managing a power station is mostly a matter of recording and logging masses of different data while looking for unusual trends before things get worse. If something does break, then the recorded data is used to find out what might have happened. This avoids having to strip down the actual power plant machinery to find out what happened, and allows the plant operators to pinpoint the source of a problem. During my time on site I saw one such episode where the electrical department used the data to go straight to the actual part of a steam turbine where there was a fault. There is so much machinery spread over such a large area, that without such a data logger the station would have needed many times the number of staff that they actually had.

The job of a field site manager is quite varied. One has to handle everything on one's own. This includes technical, business and people problems. For instance, while the power station was under construction, there were very strict union rules about who was allowed to do certain jobs. For instance, I almost caused a strike, when I was trying to debug the PCB (Printed Circuit Boards) in the CAE control cabinets. According to the union rules I had to get an official electrician to come and switch off the power to the cabinet, before I could play with my "own" electronics. As each steam turbine unit was being commissioned, I had to do a temporary modification to the system (it was still not fully ready), so that it could be used to safely monitor the unit run up. During plant commissioning all the staff are very tense as they slowly put the unit through its paces. I had to get quite familiar with all parts of the station equipment during this period. While the system was still not officially commissioned and taken over by the customer, I also had to be responsible for round the clock maintenance, just like a normal member of New Brunswick Power. I had to train the future maintenance staff too in how to repair faults. Eventually we did have a final acceptance test and the customer took over all the responsibility. I was quite proud of the system especially as it was all done with relatively little computer hardware. Twenty years later, when consultants were given the task of specifying the replacement system during the power station mid-life revamp, the consultants could nor believe how much was done by such a small computer.

In my spare time as I was mostly on my own, I did quite a bit of cross country skiing in winter and golf in the summer. Saint John was a relatively poor place then with a large population of unemployed guys many of whom also came golfing, as the city course was quite inexpensive. So I never lacked for golf companions. The countryside around Coleson cove is quite beautiful and along a spectacularly rocky sea shore. However even though Saint John is in a beautiful location, there was huge pulp mill in the middle of the city and it gave off an awful smell. During my stay I took short trips to other parts of the Maritime provinces including the bay of Fundy (world's largest tides) ,Novascotia, Cape Breton, Maine and so on. I tried various routes on my drives back and forth to Montreal, through both Quebec and the USA. When I finally returned to Montreal, all my bosses were very complimentary about my work and I think that gave me a much higher profile within the department. By now control systems at CAE had virtually shut down as there were no new projects, and almost all of the staff had left. This did give me the chance to gradually rebuild the control systems business at CAE. I was given a free hand and hence most of the technology and business characteristics of computer control systems at CAE has my imprint, probably even today long after I have retired.

The next big controls project at CAE was to supply our local power utility HQ (Hydro Quebec) with approximately 100 RTUs (Remote Terminal Unit) for monitoring its High Voltage network, with one RTU per electrical substation . This was almost entirely a hardware supply as in those days RTUs had no software. We did however have to supply some test and diagnostic software. The quality of CAE manufactured electronic systems is extremely good, and these types of RTUs are still functioning in many places, many decades after the supply. I did not have much to do with this system as it was mostly a circuit design project, and hence out of my real competence. However it led to HQ allowing us to bid on the next major control systems contract for their large James Bay Hydroelectric complex. It was winning this huge system that set CAE on its way to becoming a major player in the field of computer control systems. This period at CAE was also crucial for our flight simulator business, as it coincided with a new technology paradigm that helped us gradually overtake the competition in both simulators and controls. Let me explain.

CAE had decided to develop a completely new electronic interface system for all its products, to take into account the newly invented integrated circuit chips that were coming on to the market. This new system was called DATAPATH C to give it a catchy trademarked brand name. In fact in the future as we developed other series of electronics we kept the DATAPATH name just giving it other subscripts eg. Datapath M, Datapath50, Datapath 5 etc. Incidentally DATAPATH 5 referred to the interface set associated with the RTU and DATAPATH 50 the set used on the nuclear DCC. Many years later, another company, NORTEL, inadvertently used the same DATAPATH name for some of its equipment, so CAE sued them for trademark infringement. As I was just about the only original designer left at CAE, the lawyers used me as their star witness and NORTEL gave up their claim. The chassis backplane was made up of two connector sections. There was also a connector for the interconnection bus that connected each chassis using twisted pair high isolation wiring. The other connector sections were for the special requirements of any card and for the power supply and card calling logic. So at the back the card at the connectors looked like a "C" , hence the name! This type of bus based design was just coming into vogue, rather than individually unique card point to point interfacing. The bus controller cards could be connected into long networks of daisy chained chassis' to create large and flexible interface systems, including multiple branch segments. Even remote chassis' could be connected over a special RS485 (RS422) twisted pair serial bus called the DACBUS, that could be kilometers long, and still working like a fast parallel process interface. This new design was planned for both future simulators and control systems. In practice most of the stringent debugging took place on control systems rather than simulators as the control environment is much more demanding. This is a good example of how in a multi systems business, each type of system can help the other type of system through the cross fertilization of ideas. So in the end most of the design requirements were specified by control systems engineers, but actually more interfaces were sold to our flight simulator customers which was the bigger business.

A chassis (essentially a box) of electronic PCBs normally consists of a chassis controller card which usually has some logic to decode which card is being addressed. Then there are cards plugged into slots in the chassis that specialize in different types of inputs or outputs. The usual set include PCBs to read analog inputs, digital inputs and correspondingly for analog or digital outputs. Some of these PCBs are quite specialized and rarely used. For instance when we read electrical signals in an electric power substation, it is very important that the system record the exact sequence in which the digital inputs change. The sequence differences have to be resolved with no more than a 4 msec difference. This is a classic Electrical power network issue as the sequence of events allows the power engineers to establish how a power network fault occurred. Many control companies tried to solve this problem by designing special stand-alone SER(Sequence Of Events Recorders) complete with special trending devices added. We were one of the early suppliers to do all this in a single special PCB added to the normal DATAPATH chassis, and so could do SER actions across many substations while coordinating and displaying the results at a central control room. As time went on newer PCB types were invented to deal with special avionics interfaces for flight simulators all within the same series of electronics. As newer ICs (Integrated Circuit) chips were invented, PCBs were updated to do more and more functions. I can remember some of our older engineers quite bemused by the fact that the newer DATAPATH PCB could do the same amount of function that had required a complete chassis of electronics in the past. One element that we were quite proud of was that I think we were probably one of the first companies to have designed in a hot swap facility into our electronics. Earlier one always had to switch off power before exchanging PCB cards during debugging.

At that period of time most systems companies such as CAE had responsibility for the wiring only up to a junction cabinet where our wires were connected to one side of a row of TBs (Terminal Blocks). The actual machinery and high power equipment suppliers were responsible for the rest of the wiring to the actual equipment. Sometimes the supply of special Transducers were also included in our supply. This was relatively easy to do as we just had to buy the transducers and supply the local power to the chassis. The reason this is worth mentioning, is that there has to be a rational division of work when a customer buys equipment from several specialist suppliers. The simpler the boundary, the less likely that there will be any arguments. Another reason was that most systems suppliers were quite specialized in their knowledge and their management did not want the risk of enlarging their scope to include items that they knew very little about. Usually there were a separate group of turnkey project manager companies who were the prime contractors for the complete project and a company such as CAE was quite often several layers down as a subcontractor responsible to a higher level of contractor.

As an example, the prime contractor for the Coleson cove power station was United Engineers from Philadelphia, and they were responsible for overseeing me at site. In fact their own consultant, a very senior engineer, was most helpful to me in getting my site work done. Usually all the staff of the different companies on a work site get to know each other quite well and friendliness is the norm. So much so, I even used that senior consultant years later as reference when looking for a job (which I eventually turned down, but he did give me a superb reference!). When dealing with a very large and technically very competent customer, such as a large power utility, the utility itself was their own project manager. This was the case with most of my early projects with Hydro Quebec and Ontario Hydro. These giant utilities were large enough to have experienced staff that could do as much as we could. Often the only reason they subcontracted the work to companies such as CAE, was for financial and contractual reasons. This had a tendency to frustrate some of their expert engineers, who were only allowed to watch what we were doing and not allowed to actually do anything. So over the years some of them got fed up and approached me for a job, which they were delighted to get, even though we could not match their princely utility salaries!! This might serve as a warning to a young engineer not to get boxed into a "velvet coffin" which has great pay but leads to a boring existence. Many of my best staff came from such backgrounds, and I think they never regretted coming over.

Getting back to the earlier discussion of a technology paradigm shift, computer companies had started to develop newer and more powerful mini-computers, so we had to re-evaluate our choice of the computer on which we could standardize. This coincided with the bid to Hydro Quebec for the computer control system for their huge James Bay Hydroelectric complex. As I had just returned from Coleson cove I was given the task to prepare the technical proposal which included the choice of computer. The main choices were between SDS and Interdata Computers who had supplied the computers for our ATC project. Purely by chance I had met the salesman from the DEC company who left me some brochures about their new design 32 bit computer the VAX-11/780, which was expected to be ready about the time that James Bay would be awarded. So just to be sure, I did some technical and cost evaluations. I was surprised to see that the new computer was quite competitive in price, but vastly superior technically. So I began trying to persuade my bosses to change to the VAX. Initially they were not keen to take the risk of designing a system with all new computers, electronic interfaces and software. However I am glad I did succeed, as the VAX computer was the key reason why HQ awarded us the contract rather than to our competitor who had a much longer pedigree in the control business. As a coincidence, the flight simulator department were also looking for a new computer for a new technology simulator for KLM airlines. They took on the VAX as well, and that proved to be a real hit in the flight simulator business too. At this stage, CAE was the smallest of the three main flight simulator suppliers with barely 10% of the market. However based on developing VAX simulators over the next 5 years CAE became the largest of the three. At first we thought that it was because we had superior software, but our president told me that it must have been the VAX as that was the main differentiator between us and the competition. One should remember that in those days, large computers were considered too expensive to use in real time systems, almost a luxury. However the customers loved them, and were happy to pay the premium for a superior and more powerful computer.

The VAX computer was physically quite large, in fact it looked just like a large main frame computer. This was unlike the usual mini computers that were used in real time control. I think this computer was actually invented to compete in the large computer market, so we at CAE were unusual in trying to use it for real time computing. So for several years we had the advantage of being the only systems supplier that used such a large computer. Eventually our competitors came around to our way of thinking and the competition got hotter. The computers also got physically smaller. I mention this to show how important it was in those days to choose the fashionably correct computer in making a winning bid. The computer was the highest profile item in a bid, even though there might be other more important elements in the system. One of our principal competitors in the power business was mainly an offshoot of a main frame computer manufacturer. The reason why they were in the control business was to enable them to sell more computers. In fact their main frame computers were so large and powerful, the electronics got very hot and the systems needed special liquid cooling systems and special plumbing in the computer rooms. This greatly added to costs and reliability. With the new VAX we bypassed this issue as the computers were air cooled. An interesting psychological issue that I noticed was that less knowledgeable customers were actually impressed by the large size and complex wiring of traditional main frames, making them suspicious of the ability of the more modern VAX to compete. Eventually over many years this bias disappeared and customers began to laugh at suppliers who proposed traditional mainframes, especially the ones which needed liquid cooling. Years later I remember visiting the control center of LA power in Los Angeles which had such a computer and the technical director was quite mad at all the headaches he had with liquid cooling systems.

Having chosen the VAX, I had to include all the other elements. The operating system VMS (Virtual Memory System) was a full general purpose operating system and we had to learn to use it. It was also the first 32 bit word computer to be introduced for real time control. Another first was the use of a high level language FORTRAN as the sole programming language. All this is common today, but the James Bay system was the very first to use such advanced concepts. The Master station (central computer system) consisted of two VAX computers connected by a dual inter-computer link so that they could talk to each other. However we retained the original DCC scheme of running both computers in parallel with field inputs into both computers but with controls only sent by the main computer with the back-up computer waiting to take over seamlessly. There was the usual external watch-dog timer (WDT) that had to be clocked from each computer to show that the computer was working. If either computer stopped clocking (indicating a fault) then the WDT switched over to the other computer. The MMI (man-machine interface), called Human machine interface (HMI) in these politically correct times, was made up of display generators, printers and control panels, all set up in specially designed control desks. The control desks had normal computer keyboards for typing in information onto the CRT screens. Another innovation was a track-ball device that functioned like a modern computer mouse, allowing the operator to pan or move fast all over the screens. For years track balls were standard devices on all our control desks, as they were very convenient. It is probably only very recently that one uses computer mouse devices or touch screens. In addition there was an enormous mimic (mosaic tiled) board that showed the layout of the entire James Bay electrical network including the generating stations. Modern control rooms can use giant CRT type flexible screens for control room mimics, but in those days one was restricted to hardware fixed mimic boards.

The use of display generators was new to me as the design allowed several colour CRTs to be driven from a single chassis display generator from AYDIN. All the MMI was switched together to the control computer through a specially designed switching panel also triggered by the WDT. The electronic interface to the field electronics was through a front-end modem and serial channel system based on DATAPATH C, as was a DATAPATH series of chassis for the mimic. There were two communication lines to each field RTU. The whole master system had to have a very high reliability and availability, calculated at well over 99.8%. The new feature of the display system was the ability of the MMI to display schematic colour one-line diagrams in addition to the usual text based items. This was done by incorporating a specially designed set of icons into the display generator hardware memory. Using this technique, full displays could be called up within 1 second, so the system was very responsive to operator input. Initially I did not understand why that was important, but later I realised how important it is for a system to be fast and responsive to operators. During electric power system emergencies they should not be harassed by slow control system responses that divert their thinking from analyzing the actual external electrical system faults. There are several important design goals in evaluating a computer control system. One is the speed of response to operator commands, another is speed of system response to field changes. This fast response feature was a crucial evaluation item in choosing the CAE system over the opposition. There were three control desks with three CRTs each, so one could call up several diagrams at the same time and the response had to be the same with no deterioration. I know it sounds trivial, but it is actually quite difficult to achieve especially as the number of CRTs increases.

As this system was a prestige project for HQ, they wanted the control room to look its best when visitors were shown around. So the room design and all the furniture was under the control of a special HQ architect, who insisted on a specially attractive design for the control desks, with the CRTs and panels neatly arranged together with the appropriate telephone and wireless equipment. This introduced me for the first time to sub-contracting. As the desk was made of steel with a special blue non-erasable coating, we could not build it easily in our factory, so I had to go out for bids to specialist cabinet makers. Eventually we found a jovial steel specialist who did a first rate job in making the desks, so we gave him the task of building the Electronic cabinets as well. The mimic board (tiled layered mosaic) also had to be sub-contracted. This other sub-contractor was quite a small company and our contract was a big project for him, but he did a superb job. In fact our sub-contract allowed him to grow his company and it became quite famous. Recently I noticed that his company has made the outside display on Times square in New York! It can be interesting, working with small sub-contractors. They live quite close to the margin and are very competitive, and usually very responsive. So I have learned to nurture them and get to know them well. It always pays off as they will make a special effort for you when project schedules get tight. This is unlike the work done in our own large factory where the manufacturing bureaucracy responds very slowly.

The field system consisted of a large RTU at each electrical sub station and at each generating station. The James Bay system was spread out over a large geographic area in sub-arctic northern Canada. The total number of signals to be monitored and controlled was well over 18,000 making this system one of the largest ever to be controlled from a single master station. This very large database and the equally large number of single line display diagrams required an efficient software design in the VAX and the front end processors. Once one uses a general purpose processor such as the VAX, one has to solve the problem of external interrupts efficiently. In the old days the specially designed process computers such as the GEC 4080 or the M2140 had specially designed I/O processors for this. So we had to design a new type of front-end system. This front-end processor was a CAE special design computer (one of the first ever bit-slice micro-processors) called the DUSC (Dacbus -Unibus Smart Controller). The Dacbus was the RS 485 bus that connected all the CAE designed DATAPATH C chassis', the UNIBUS was the DEC computer (VAX11-780) company's own bus used to connect computer like devices together. Hence we needed this new design controller to connect the two types of electronics together. It just so happened that the micro-processor in the DUSC was so powerful that it had plenty of capacity to run all the front-end and communications software for the master station as well. So we did not need another separate front-end conventional process computer. This turned out to be a systems advantage for us, as most of our competitors had to provide much larger and more expensive conventional full computer type front-end systems.

Systems design is eventually a competition to come up with the most powerful AND the most cost effective system. Communications software requires very fast responses to random interrupts from the field, so it was not reasonable to use the main computer. Large main frame type computers do not usually have good interrupt capability and can take a long time to change software tasks according to high priority interrupts. Microprocessors on the other hand are easy to program to receive interrupts and change software tasks quickly. Especially if there are many serial lines coming into the front end systems, they are also easy to expand simply by putting in more microprocessors as the number of serial lines increases. In earlier systems, fixed communications logic was actually hard-wired into the electronics. Front-end computer software was a new technical approach, which enables one to change the remote control protocols easily and so one can handle different RTU types from the same basic hardware front end.

The system time was received from a special IRIG-B system. This was provided by HQ and is important so that the control system and the electrical network are both time synchronised. On this system the RTUs were all hardwired with pre-allocated proprietary message protocols. So they responded very quickly. The long distance communications was done over the HQ communications network that covered the whole of the huge province of Quebec. It was a microwave high speed system. There were redundant links to each RTU, with our modems to get our messages onto the system. The line speeds were 2400 BAUD which was considered very fast in those days. As is normal, the RTU electronics actually interfaced with the field devices through wiring done on a set of TBs at the back of the RTU cabinets. These cabinets were huge because of the large number of signals. Our cabinet hardware designer got carried away and specified un-necessarily high standards. In fact when one of our executives first saw these huge structures on the test site, he laughingly suggested that they had been designed to foil an attack made with hatchets and welding torches!

When building a system where all the various elements are miles apart from each other in the field, it is best to do a thorough system test in the factory where everything is close by. Otherwise it requires armies of staff spread across the various field sites. In fact HQ insisted that they would not allow us to ship the system to the field unless every single snag (error) had been fixed first. Superficially that sounds reasonable, but it is actually very, very difficult to achieve. The final system testing alone took almost two years, and eventually this system was probably the first large real time software system ever shipped with ZERO snags. It is fairly normal for customers to accept a system with some snags as long as the snags are not critical snags. Otherwise it would take a very long time to wait for perfection, especially as large computer control or simulation systems can have over a million lines of software code. This system has since run fault free for many years. To give an example of how good it was, several years later I learned the following. HQ had decided that they wanted to move the control center to a town 300km away as it was getting difficult to get their operations staff to work in the remote James Bay region. So they moved half the system first to the new site, leaving the other half to keep the control system running. Then they merely switched over to the equipment at the new site, followed by moving the remaining equipment there as well. During all this effort the James Bay controls were never compromised for more than the 1 second usual switchover time, just as if they had merely done a switchover locally!!

At that time to curry favour with our French speaking clients, we decided to run project meetings directly in French, even though many of us were barely proficient in the language. We even tried to find equivalent French words for technical items such as memory overlays etc. At first our clients listened politely but looked confused. It turned out that they actually used English technical jargon just like us and did not understand our fancy made up French equivalents! In the end the system was a great success with our other customers too. Several years later I took some Chinese clients for a visit to Northern Quebec to see the James Bay region and the various hydroelectric facilities. The HQ staff were so proud of their system, that by then they thought that they had developed the system themselves and no one recognized me as an actual designer. The James Bay hydroelectric complex is truly impressive. Currently the total generating capacity is about 15000MW from 5 power stations on the La Grande river, deep in the arctic region. I believe plans are still there to increase this capacity by diverting other rivers into the La Grande and thus increasing its flow. A major benefit of doing the controls for such a large hydroelectric system (largest in the world at that time) was that we got the reputation of being the most experienced supplier of control systems for giant hydro systems. As a result, for several years after that, other large hydro-electric complexes came to us looking to buy similar systems. Even though this system was very fast and powerful, the actual hydro-generation software applications were rather simple. They consisted of mostly managing the allocation of generation to each actual hydro-generator by a balancing algorithm, so that water use was minimised. All the generator units were identical so we could assume that there was no efficiency advantage in preferring any particular generator. The electrical system was a modern high voltage 735KV system that transmitted the power down three power corridors to Montreal around 1000km away.

These types of system are in such remote and isolated places that there are few staff to manage and fix things, so equipment has to run for years on end with no failures. In addition the environment is rough and very cold. CAE RTUs are very well made and very reliable. Some years later I was on a site visit to Churchill Falls hydro-electric station as they were interested in buying a computer control system to replace their hard wired system. This power station is also in the arctic region of Labrador (in Newfoundland)and also supplies power to the south via the HQ system. As this station sells power to HQ, they had put one of our (CAE) RTUs to interface with the power network there. This RTU had never been opened by their staff and had functioned flawlessly for years to the point that no one dared to tamper with it. So they were delighted to see me and asked for more details about how it worked. Incidentally it is good idea to never open up cabinets and fiddle with electronics, as long as the equipment keeps working. Many problems only arise when someone starts to open up electronic cabinets! This hydro-electric station is also huge (around 5000MW). Unfortunately the client was reluctant to move to direct computer control and settled for only specifying a data logging system. As our prices were too high, another company won that contract. Flying into Labrador is quite a long journey with a change of flights in St John Newfoundland, where one changes to a very small plane for Goosebay and Churchill. All northern Arctic travel is heavily weather dependent and we nearly got stuck there due to bad weather. It would have been very boring to be stuck out there as there is little to do. The airport itself is just a wooden shack with no one usually there. At the dam site there is a very small village with a few houses for the power station workers and a small hotel with associated shops.

However in Canada, we did win another large hydro-electric control system for Manitoba Hydro at their new station on the Nelson river basin north of Winnipeg. This was for their Limestone station also in the far north. By this time we had become famous for such systems in the USA, Venezuela and China, so the client was happy to work with us. Clients were generally confident enough to give us both the computer control system and the hardwired fallback system with all the required hardware panels wired with meters, switches and transducers. In earlier systems the power plant engineers always insisted on a separate, completely independent, hardware only, simplified manually managed backup system just in case there was a problem with the computer system. Usually this back up system was supplied by the turbine or generator manufacturer who understood the machinery very well. This Limestone system of course had the usual dual computers and control desks as well. It was a similar system to one we had designed for Geheyan in China and I will describe that later.

Over the years, I visited almost all the provinces of Canada trying to win projects at the different power utilities. We did design bids for systems in Novascotia, Quebec, Newfoundland,New Brunswick, Ontario, Manitoba, Saskatchewan, Alberta and British Columbia. It was interesting work and gave me a good overview of Canada. Each province had its own power utility, often called "hydro" as most of them had started with just hydroelectric power and then gradually moved to other fuels like nuclear and coal. Unfortunately they never gave us any advantage for our Canadian background, they awarded contracts solely on price. Hence we had a difficult time persuading potential clients to pay a premium for our more powerful systems. In the end we won our share of the very largest systems, but none of the smaller ones.

One of the biggest computer control systems that we won, was the project for all the regional control centers for the HQ high voltage transmission system. There were 7 systems and 1 training system. At the time it was the most important and largest such project in the world and the competition consisted of all the major suppliers from Europe and USA. HQ had a large and competent systems staff, so they understood the advantages of buying the most powerful system. Their specifications were not only very stringent, they had an actual experimental test as part of the evaluation process where they actually measured the real time performance of the proposed computer systems. So in addition to the usual paper design for the proposal I had to set up a small programming team to demonstrate the simulation test which was witnessed by the client. Once again a major advantage was our use of the VAX computer and our expertise in fast software developed for James Bay. These regional control centers were several times larger than James Bay with the largest center using up to 40 CRT screens spread among several control desks. The largest signal database was around 40 000 points, so there were many front-end chassis to cope with the large number of serial channels. Our competitors all offered very large main frame computers, with separate front end computers, but still could not match our expected performance. Another advantage for us was that as we had already developed some similar software for James Bay our price was significantly lower too. Our inter-networking technology between the various regional computer systems, based on DEC's own networking software was also the best available at that time.So we won and it turned out to be a huge success. It was also crucial for the CAE company as a whole, for the contract award occurred during a period when the larger flight simulator market was in a slump. So the cash flow from this huge contract kept the company afloat during a crucial period. It was quite eerie to see the company factory test sites full of only computer control systems and no flight simulators. There was a separate contract let for around 400 field RTUs (one at each high voltage substation). HQ themselves had planned to write special applications software for all these systems.

As was usual with HQ, the systems had to respond with very fast requirements for inputs (analog and digital) and in huge quantities without losing a single event. Many years later after all the systems had been installed and were functioning, there was major sunspot generated magnetic activity over North America. Quebec is an enormous province. It is much larger than France,Germany and UK combined, so these regional electric networks had power transmission lines several thousand kilometers long and so susceptible to induced currents from solar magnetic storms. This solar storm led to the entire HQ network shutting down as huge electrical currents were induced into the long distance power lines. This was the toughest possible power system disturbance imaginable and our control systems came through with flying colours. Not a single electrical event registration was lost. However the printers could not keep up with the enormous number of alarm messages, so the system operators could only wait several hours for all the thousands of messages to come through. Eventually this recording of the electrical system disturbance was analysed and that is how the utility was able to diagnose the magnetic storm induced power failure. Many years later there was a major ice storm that shut down the entire electrical power network around southern Quebec. This emergency was also successfully managed centrally from these control centers. I can remember being without any heat or power for well over a week in the middle of that cold Canadian winter. All the HV transmission lines around Montreal fell down after being loaded down with heavy ice. It was quite an experience and goes to show what a difficult job a power utility has, to manage in such a rough environment.

As we used a general purpose operating system (VMS ie Virtual Memory System), we had to modify our software design approach. This was an early design based on using the idea of independent processes (programs) for each major function. A software process is a safer way to isolate a program as it includes all its parameters and resources separately. If there is a fault in the software it can be isolated from the other processes. This is a useful safety feature and a good way to debug a complex system. However one must chose carefully which functions deserve to be full processes and which can be basically just sub-programs within a process. With this type of design, one can gradually introduce new processes into the system, rather than the older technique where pretty well all the programs had to be introduced together during system testing. In earlier systems the computers were so slow, we had to invent home-made executives that ran optimally and did not miss real-time interrupts. In this design the front-end computers (DUSC) had to handle most of the external interrupt traffic, leaving the VAX computers to handle the MMI , applications and databases. The MMI real-time effects were directly handled within the AYDIN display generators. Again this was a big advantage from the old days where we had to handle these real-time effects within the old mini-computers. For several years these AYDIN type semi-graphic raster display generators were the only designs good enough for control rooms, as full graphic vector generators would have been too slow. The scanned in data was kept in a shared memory area organized just like the flight simulator shared central data area. The power equipment parameters were put in using a new database editor, while the display pictures were built with a new display compiler. For the various reports and alarm monitoring activities we had a set of specially developed programs.

At around this time HQ also awarded us the contract for computer control of a new gas turbine peaking power station outside Montreal at La Citiere. We had to use the smaller 16 bit PDP11-70 computers for this design as we had to have a lower price. However the rest of the design was similar to the larger systems with AYDIN display generators, DUSC frontends and DATAPATH C electronics. This system needed to be too fast to use the RTU type field electronics, so we just extended the DACBUS all over the power plant with separate DATAPATH C electronic chassis' at each gas turbine. As the power plant had been designed to include upto 32 gas turbines, the station was spread over a huge area just within a 735KV high voltage substation. This DACBUS created the speed effects of a parallel field electronics system, but was actually a serial bus system and saved a great deal of wiring. Another advantage of this design was the effect of avoiding earth fault issues. These could have been very severe as the DACBUS could have received high induced voltages from the surrounding 735 KV High voltage switchgear. We solved this safety issue by allowing the DACBUS to float, unless one opened a chassis cabinet, when this opening effect automatically grounded the system.

During this period, Ontario Hydro OH, the power utility in the neighbouring province of Ontario, was in an expansion mode. I have already mentioned the DCC systems for the nuclear stations at Pickering (near Toronto) and Bruce (Lake Huron). The next nuclear station that they planned was a much bigger plant at Darlington (lake Ontario). This time Ontario Hydro engineers had decided to do all the applications software for the station, rather than give it to AECL, who were the nuclear island supplier. As is usual, a new team meant a new design. We did win a similar contract to supply the DCC, but it had to be to a new OH specified design. Instead of Varian computers it had to be based on PDP11-70 computers, DUSC front-ends and SER recorders, RAMTEK display generators and DATAPATH C field electronics. OH refused to consider VAX computers as they were considered too risky as they were a new design and hence risky for use in the nuclear industry. The DATAPATH C electronics we used were essentially a specially designed set of PCBs exclusive to OH and nuclear stations, rather than the same cards previously mentioned. The RAMTEK displays were a new full graphics design from a specialist company based in California. This eventually introduced me to the flaky world of Silicon Valley. Stuff from the Valley was great in terms of new advanced designs, but they never properly finished what they started. For one thing there was a tremendous turnover of staff at these small companies. In the end as the systems supplier, we ourselves had to fix the display generators and modify them adequately so that they would work reliably for many many years. These nuclear DCC systems had to go through a very stringent factory test with zero faults over a period of 3 months. Even one small fault meant that the whole factory test had to be completely restarted. I think it took several years to finish these systems, but in the end it all paid off and they are still running very well at the Darlington power station.

As OH were a very innovative company, they had also introduced a new PLC (Programmable Logic Controller) the OH180, to handle secondary inputs and controls all across the power station. They had invented the device themselves and greatly over specified it, so it was great technically but very expensive. In the end it turned out to be a typical "white elephant" as they could never get anyone else to buy it. Even their own fossil power plant division refused to use the device as it was too expensive. As a desperate measure they asked me to take it on board as an extra device that CAE could offer our customers, but it was no use, none of our customers were interested. It is a good cautionary example of the folly in over engineering and over specifying designs in a competitive market.

This mention of PLCs reminds me about the fashion and habit that pervades even the supposedly rational world of electronic engineering. Today there is not much if any difference between a RTU and a PLC. However they were each invented for different industries. The RTU probably came from the electrical industry especially for use in substations. The PLC came from the industrial mechanical factory world. As I was mostly initially involved in HV electric power, I naturally gravitated to the RTU concept. A PLC is really first rate as a simple controller using ladder logic that is easy for less trained shop floor staff to manipulate. So the simplest PLCs do not need any extra software support. I first came across the PLC when doing the bid for the replacement crusher control at Kiruna in Sweden. The client had changed their ideas and wanted PLCs rather than RTU like outstations for their next system. As we lost that system I heard no more about PLCs for many years. Once when visiting BC Hydro to talk about control systems for their large hydro-electric stations on the Columbia river, I tried our RTU type approach with their engineers. Their chief engineer was a firm PLC guy and so I had to give up. Even in the marine environment PLCs were quite common. So I tried to be flexible and offered what I thought the client preferred. A good analogy in today's world is the choice one has to make between Wintel or Apple computer environments.

We at CAE had our own problem with our electronic interfaces too, as we also tended to somewhat over design equipment as well. The advantage was that our customers loved the stuff, but were not happy with the prices. A further cost issue that we had was related to the fact that our factory had been set up to build custom systems and was generally somewhat less competitive than many of our competitors who built electronics on a production line basis. This issue gave me lots of trouble with my bosses. I discovered that it would have been a lot cheaper to offer RTUs from our competitors when designing our systems, but our president would not allow it. He insisted that it was my job to teach the factory and circuit group how to be competitive, and he refused to let me use any outside equipment. Years later I came to realise he was correct (as he usually was). If you don't learn how to be competitive in making things, you will soon hollow out your company and lose all your ability to win systems. To be truly successful in the systems business one must make stuff not just buy in. Only then can one properly understand the subtleties of systems performance and reliability. It is a risky business buying paper designs from companies that don't make anything, but just buy in. Whenever there is a problem they don't know how to fix things even though they may be contractually responsible for systems integrity. So I had to try and learn exactly how our factory worked.

What I discovered was that the actual wire-men who put the equipment together were extremely efficient and competitive as were the PCB fabrication facilities. The problem lay with the manufacturing control method and process that tracked (on paper) where everything was at each stage of manufacturing. I think the method chosen was taken from the process that large production companies (such as auto-mobile manufacturers making millions of identical parts) use. This was inappropriate for us, as we rarely made two similar systems,. Each system was mostly a custom system,hence a "cottage industry" approach would have been more sensible. This did not seem to matter too much when making flight simulators as the market was something of a restricted oligopoly. However it made my life hell in the highly competitive control systems market. The factory also seemed to be very inward looking and seemed to think that at CAE we only built flight simulators. I remember a hilarious meeting where the factory director was checking progress by asking when various items were due on the test site floor. As he was asking when various simulator parts were due he suddenly came across the plan for the Limestone Manitoba hydro power plant electronic interface. I remember him asking when "the Manitoba cockpit" was due on the test site as if it was just another flight simulator! There were hoots of laughter all around.

One of our major failures was due to some really bad luck. At this period, our marketing department had discovered that most of the Canadian provincial utilities were planning to buy new EMS systems. In view of our successes in winning large hydro power control systems, we decided to enter the relatively new market for EMS systems too. An EMS (Energy Management System) is a computer based control system that is used to monitor, control and manage high voltage power flow across a region. This involves the usual SCADA (Supervisory Control And Data Acquisition) system that will use electronics to monitor and control the physical high voltage power equipment all across the region. The extra new wrinkle was that the computer system would also run a series of power applications software that simulated and calculated the actual security status of the HV power grid, using very new and sophisticated power systems software. Of course the power utilities were only interested in dealing with those companies that actually had the experience of having provided this type of system. At that time there were only about 3 or 4 companies that could claim the requisite experience. So we had to come up with a strategy to overcome our lack of this specific experience. We hit upon the idea of teaming with an already experienced utility who could cover our weakness. Ontario Hydro had themselves designed and built their own EMS for the whole of Ontario and wanted to get back some of their investment by selling their software to another utility. So we joined up with them and tried to propose this partnership on EMS bids. ie.OH to supply the applications software and power system expertise, CAE to do the rest of the system, including being the prime contractor.

So after some preliminary work, we went on a trip around Canada to talk to potential provincial clients about our approach. Unfortunately it did not work, though Saskatchewan Power did allow us to bid, but we lost anyway. I discovered that having OH as a partner turned out to be a disadvantage, as the smaller provincial utilities were worried that the huge OH would end up dominating them, so they turned down our requests to bid as a team. Eventually BC Hydro allowed us to bid on our own using a commercially available power system power security software application package that we had bought and integrated into our system. I think their engineers were more impressed by our electronics, and their American consultant thought that a Canadian supplier would have been a political plus. We won the competitive bid and were on track to sign the contract when there was a political fallout in BC and the president of BC Hydro was sacked and as a consequence the bid was cancelled. This gave me some problems as I had hired some new staff and was ready to start quickly. In the end I got some other projects so I was not put into the embarrassing position of laying off staff.

Newfoundland power, New Brunswick Power and Nova Scotia power, wanted a completely standard system with no special customising. This was a disadvantage for our bids as we were really only competitive when bidding custom systems. So we lost these systems too.

However all was not lost as we persevered and kept learning more about HV power systems,eventually developing our own series of power applications which caught the attention of American and Chinese utilities.So we were able to finally become part of the small elite group of companies that could bid full EMS systems. Many, many years later our own local utility Hydro Quebec decided to create a new EMS control center with the most advanced power applications. It was at that time the most prestigious system bid and was heavily fought by international competitors. We won, partly because we had the best technical bid, but also because we were local and were favoured by the provincial government who was the owner of HQ. So we started the joint design phase with HQ. Unfortunately there was an election in the province and the government was replaced by a new political party in power. This party decided to cut spending and the HQ control system was cancelled. So once again we lost an excellent opportunity to develop a great system. However in the end some years later HQ did award us the contract for a new less expensive complete EMS system.

In later years the North American electric market was deregulated and wanted EMS systems that included commercial software such as price auctions and analysis for electrical block transfers. Ontario Hydro asked us to bid on such systems, but in the end they just replaced their control center by updating their own original design at their new control center building. For some reason we were never successful with Ontario Hydro's Transmission or Fossil businesses, but were quite successful with their nuclear business. We did have many examples of teaming with them for foreign business as fellow Canadians, but it did not lead to anything, though I got to know many of their senior staff over the years. Eventually we did get DMS (Distribution Management Systems) systems for the distribution electric systems of the cities of Mississauga, Scarborough and others in the Toronto area, after we had pioneered such DMS systems in the USA.

As we at CAE were essentially a builder of custom designed systems, it meant that we were constantly on the look out for unusual or new types of systems. For a control system to be a good business fit for us, it had to be at least as large as a typical full flight simulator with the work required to be an equal split between engineering and manufacturing. That was how all our financial overhead allocations were based. In general most process industry type systems did not meet that criteria, but it did not stop me looking. One of my colleagues had done some research on factory automation at university. It also was a fashionable topic at around the time he joined us, so we began to actively look for a target factory automation system. We found a project just down the road in Montreal itself at Pratt and Whitney. They were planning a new aero-engine manufacturing factory and they wanted it to be set up with the newest automation technology, robots and all. So we spent a great deal of time coming up with a design and a proposal, but eventually lost to a much smaller but more specialised company. It turned out that this market was not a good fit for us. It also turned out to not be a particularly large market segment either.

Another foray into industrial automation happened because we had very close ties with an associated company that specialised in making machines for the forest industry. They persuaded us that there was a potential market for full lumber mill automation as Canada had a huge forest industry. As an outsider I had assumed that wood mills were simple factories with not much automation. I was quite wrong. They use some very sophisticated laser based machines to optimise the cut of wood by analysing each log very carefully to maximise the use of the wood. We spent a couple of months out in the Vancouver area visiting lumber mills and seeing if there was a possibility to automate a complete mill. Unfortunately at the time the forestry industry was in bad shape and no one had any money to spend on automation. So that idea died as well. As Canada is very big in mining too, we tried some half hearted attempts to look at systems for mining automation but once again there was no fit. During my period looking for work in military systems, I was once invited to the military base (Valcartier) outside Quebec city to see if we could commercialize a military mapping system developed by the government labs. Nothing in it for us, but I did get an invite to the officer's mess for lunch. Unfortunately it was really hot (yes, Eastern Canada in summer is boiling) and I had not bothered to pack a necktie. Fortunately someone lent me one, or I would not have been able to enjoy the hospitality of the Canadian army. I also made lots of visits to our Canadian NRC (National Research Council) laboratories in Ottawa, as we had close contacts with them. They were looking for a commercial company to take over some of their technologies in police control centers, artificial intelligence, medical devices and so on. Nothing much came out of this, though for a time CAE did work on an artificial heart device. Other Canadian research institutes such as CRIQ in Montreal also had nothing much for us.

Other types of systems that we actively pursued were large oil and gas pipeline automation projects for Transcanada pipelines and Trans Quebec and Maritime pipelines. Once again the required SCADA systems were too small for us to be competitively successful, though we spent a great deal of time learning about gas calculations and other specialised technologies related to gas transport. Generally speaking, process industry systems, while sophisticated in their own way, did not need our particularly fast and flexible electronics. These industries were themselves in very competitive markets, very price conscious and not willing to pay extra for our larger systems. They also had a whole slew of experienced suppliers who were more competitive than us. So I generally gave up expecting much from outside the Power, Naval and Air Traffic management industries. Years later, even as an independent consultant I tried again looking into and studying oil refinery, pulp and paper,light rail automation and mining industries, but kept hitting the same blockages. They all required different types of system companies with slightly different strengths, even though in principle we could have built systems for any of these industries. One strange request came from an oil company that wanted us to develop a training simulator to train drag line start-up operators for an oil sands project in Alberta. We thought it would be too expensive for them, but at first they did not seem to care what it cost as they expected the cost to be easily less than any repair costs from operator errors. Unfortunately after my analysis I thought such a system would not be ready in time for the start of production, and the operators would still have to go directly onto the live equipment. So the bid was cancelled as it would have been pointless.

One area where CAE did successfully get into smaller systems was the construction of smaller flight simulators often called cockpit procedure trainers and simple simulators without visuals or motion systems. These were used to train pilots in stages before getting to full flight simulator training. Even here we usually only managed to be competitive when a client bought a complete suite of simulators. It all goes to highlight the fact that different companies can be competitive only in systems of different sizes. It is a rare company that can do everything.

7. South American cities and jungles

The GURI dam is a huge electric power facility deep in the middle of the jungle in Venezuela. It can be viewed as part of the immense Amazonian basin. The dam is on the Caroni river, a tributary of the mighty Orinoco. This remote part of Venezuela was allocated to a specially created state owned company mandated to develop this area, called EDELCA (Electrificacion del Caroni). The dam is so big it supplies the majority of electric power required in the whole of Venezuela. The hydroelectric complex generates about 10,000MW from two power houses on the dam. The newer power house 2 at 7000MW is among the largest single power stations in the world,

At the time that we got involved, we had just finished the James Bay hydroelectric control system and were looking for new work. Our reputation had spread and Harza the consulting company for GURI came to visit us to encourage us to bid for the new control system that they were planning for the complex. At first our bosses were a bit reluctant to design and build a critical system so far from the regions where we were comfortable. However Harza insisted we at least visit the site and assess it for ourselves. So we went for a ride out to the dam site. For me it was a real eye opener. The region of the dam was hilly and deeply forested, very pretty but terribly hot. The civil contractors of the dam had built a small township for their construction crews, complete with club houses, hotel and restaurants. They even had a beautiful 9 hole golf course, where I later honed my driving skills during site visits. There was an existing old power house with 10 units totalling around 3000MW that had been running for sometime in the past. To develop the new 7000MW plant they were raising the height of the dam to supply more water to the new turbines. The 10 new units were each capable of around 700MW and the unit contracts were very prestigious for the suppliers as it enabled them to develop these huge turbines and generators. This meant that the control contract was also a prestige project and reserved for only those companies that Harza thought were the best world wide. We came in because of our experience on James bay. Later the consultant told me that one of the key reasons for eventually choosing us was our experience with VAX computers, which was then unique in the world

The required design was very innovative for that era. There was the usual dual computer master station , but they also required a backup system based on another single computer control system for each of the two separate power houses. There was also a requirement for an independent computer driven RTU at each turbine unit. The master station also had to have a separate computer system to support the maintenance management system for both power houses. They also decided that they needed a separate training system for the operators who would manage the station. Our design was based on dual Vax 11-780 computers for the main master control system. There were several control, training and maintenance desks with Aydin display generators, plus the usual array of printers and mimic board. The maintenance and training computer was a third VAX computer that was connected to the other two through a new design that we pioneered using a duplicate shared memory system. In this fashion the three computers could seamlessly take on each others tasks if necessary. The usual front-end DUSC computers were linked by serial buses to the separate power house computers (Intel 8086) down near the station turbine hall. There were additional links to each separate RTU as well. So essentially there were three degrees of redundancy ie. redundant master station, independent power house and independent RTU. There was also a networked independent switchyard computer system up in the nearby hills for each of the 800Kv and 400Kv switchyards.

This was the first time that the power industry would use a computer software based independent RTU. The intent of the station consultant was to use less expensive normal mini computers eg. DEC PDP 11-34 at each RTU. However the IC chip industry had just introduced new micro-processor chips that we had assessed. So I bid the use of Intel 8086 micro-processors for each RTU computer and each station computer. I had calculated that the processors would have been powerful enough to run all the SCADA software locally. This proved to be quite crucial in our finally winning the bid on both a technical and price basis. There were around 10 international bidders (Japanese, Swiss, Swedish, American, German) for such a prestigious project so the battle was a tough one, but in the end, price, the VAX computer and the innovative computer network triumphed. This victory put us in again in the front ranks of control systems suppliers, especially for the largest hydro power stations.

The choice of Intel was mostly chance. At the time Motorola also had developed a microprocessor family of chips. In fact I think they may have even been more appropriate for use in real-time control, but somehow our circuit design department chose Intel, which was fortunate, as years later it remained in the lead and Motorola is no longer so big in IC design. There were other bids where Motorola chips cropped up but generally we stuck with Intel.

The software was very advanced (for the times of course) as well. We had to develop independent 8086 based SCADA software based on Intel's PLM language and the RMX operating system.. We were lucky to cover the development costs on this project as it gave us a unique RTU for other markets. I expect this RTU was probably the first entirely software driven RTU in the power industry. At the master station in addition to the usual SCADA software, we had to develop a unique maintenance management software package for managing the large amount of equipment that makes up such a large power station. This involved creating one of the first independent database systems to manage the equipment data and the displays that were used to interact with this data. In those days database systems were expensive and used only in the largest commercial systems eg. banks, and were unknown in real time industrial systems. For sometime into the future, we tried to make this one of our advantages when bidding for other large hydro plants. However in the end the computer industry came up with standard database systems that were relatively inexpensive and our proprietary system was discontinued.

Another innovative first was the development of an operator training simulator for which we had to develop a complete software model of the power station and its associated electrical grid. We had to develop an instructor station with software to manage the training session, and insert faults to see how the trainee operator reacted. The way this simulator worked was that the computer running the simulation software interfaced through the shared memory data link with the normal standby computer running the actual control software. This control computer had its own simulation operator console for the trainee to use. This was an industry first. In the past operators were trained on the actual control equipment during quiet times. However it was too dangerous to train with realistic problems when using the actual power equipment. As CAE was also a simulation company we had the skill sets to program realistic software models of the power equipment,using first principle physics. The design consultants of the power station had all the construction and design details, which allowed us to then use that data to prepare the models. As this was a success, we began offering a built in training simulator as an extra option on many of our control systems bids to give ourselves a unique advantage. As new power plants and other machines get more and more complicated, I notice many customers are beginning to specify a training console and simulation as an extra option for the most advanced systems. It is a good example of how newer systems include features from previously successful implementations.

The control system also had to directly manage the allocation and control of generation to individual turbo-generator units based on efficiency and economics. So we invented a unique Hydro based AGC (Automatic Generation Control) and (ED)economic dispatch software. The AGC mostly involved straight forward load frequency control of the GURI grid. This used a closed loop control strategy that appropriately changed the amount of generation to match the load while keeping the frequency steady. The Economic dispatch involved allocating generation to units dependent on which generators were the most efficient, while allowing each generator to be balanced with enough spare capacity for any suddenly required increases in demand. The regional control center EMS system was based in Puerto-Das near the end of the Caroni river and we had to have interlinked computers with their system too. Even though that system was theoretically supposed to instruct GURI as it was a higher level system, because of GURI's enormous size and importance to Venezuela, GURI was usually allowed to run while independently managing the AGC. Individual generator sequence control logic reminded me of my times in Sweden with the crusher machinery. There was lots of safety logic required for automatic control and monitoring of such huge combinations of machinery.

In a sense this was our first real foray into the new field of EMS (Energy Management Systems) and GMS (Generation Management Systems). These were new technology concepts just being introduced into the power utilities in North America. The main drive to introduce these systems into power network transmission control rooms, was to safely optimise power flow. This was a consequence of earlier failures of the North East American transmission grid which caused large power outages. This had disrupted the whole NE of USA some time earlier. Before this time computers were considered too risky to be used for such closed loop controls, but modern computers were getting a good reputation for reliability. On the strength of our developments on GURI, we tried to win some of these early EMS projects, but it took a long time before we were successful.

Eventually GURI was a great success. Our site manager decided to have a party to celebrate and invited almost everyone in the village of Guri to attend. He had a budget only large enough to pay for either the best food or the best band. He chose the band. It was so loud, even though I left at midnight, I could hear the trumpets all night from over a mile away. In the years that followed, this customer came back directly (without re-bids) to us several times for major upgrades to the computers and software.

Once our name was known across South America, we were put on various bid lists for several projects in Venezuela, Ecuador, Colombia, Mexico, Argentina, Chile and Brazil. Only a few bids were successful as the competition was usually very fierce and our speciality generally was for only the biggest and most complex systems, which were rare. Finance was also a problem as the utilities rarely had any extra money for complex systems. As large parts of the bid documents, especially the commercial documents were in Spanish, I picked up a smattering of written Spanish. It is quite easy to manage if you know some English and French. The really superb translators were some of our French Canadian software engineers. They picked up the language so quickly that I even saw them debugging their software on the GURI site in the Spanish language directly together with the client's staff.

A bid for a large system, being a very complex engineering task of course has to be done to a fixed time line. Very occasionally the client may give a time extension for bidding, but that was quite rare. The client would produce a large document that described in great detail the exact technical requirements that they wanted in the contracted system. The way the proposal engineering team had to work was to reply in great detail to each and every requirement showing how our design met that requirement. All the main engineering overview designs have to be done on the actual proposals, so it can be a major engineering effort in itself, usually by the most experienced engineers. This design is then sent to other more specialised departments such as manufacturing to accurately price this design for both labour and material. The engineering hardware and software departments had to provide accurate estimates of the effort required and produce a working level time schedule for the project. In parallel the commercial departments analyse and reply to the commercial and financial requirements of the bid. The final price is vetted by senior company executives after adding on various overhead charges. These bid vetting meetings can be long and acrimonious as there is a bit of guesswork as to what the winning bid price should be. Initially, as we were quite new to the business, we had very little knowledge about what prices our competitors might bid, but eventually after years of experience and analysis I got fairly good at guessing very close to the actual winning prices. This effort is a bit like working in an intelligence service. It involved using all the publicly available information after the bids closed, plus understanding our competitors systems and equipment. Usually this initial bid effort took about 3 months. Then there was an extra 2 or 3 months during which the client sent out questions to further understand our system, followed by them choosing a short list of final contenders.

Finally there was a series of one on one meetings where the client got familiar with the engineers who would work on the system, if chosen. This was a crucial meeting as the client had to build up trust with the potential supplier staff. So it was not unusual for the bid team to be involved in the bid for long periods up to a year or more. This process was extremely expensive and our management would not allow us to bid unless we could make the case that we had a good chance. In fact usually clients did not even allow a company to bid if they could not pre-qualify by showing relevant experience. There were other problems when the prices were opened, often leading to long political fights where the losers would try to aggressively get the winners disqualified. So the marketing team had to cover all the bases ie. price, technology and politics. We could be up all night in the field juggling our offers and negotiating on the spot while keeping in touch with our bosses in Montreal by whatever telephonic means were available then. There was no Internet in those days and the best method sometimes was through poor fax or even teleprinter communications.This was a major effort and the marketing team leader had to be prepared to cover everything on the spot, though that made it a very interesting job for him. In the international market, one also had to have a good feel for various international cultures something some of our other American and European competitors did not really appreciate. As Canadians we were at an advantage without any colonial or hegemonic baggage.

The only business area where the bidding was straight forward and quick was for the commercial flight simulators as the bids had been standardized. Some bids never even came to engineering as the sales department handled everything by themselves. This of course was not possible in any of our other business areas ie. Power control, Marine control, Air Traffic Control, Power simulators and Military simulators, all of which required heavily detailed (and expensive) bids.

One project that we chased very seriously for several years was the EMS system for the power company HIDRONOR based in North Patagonia in Argentina. The way we got in on that bid was through one of the design consultants to GURI, who had analysed and liked our system. This consultant was a world famous figure who was well known in the power industry. In fact he had pioneered most of the power flow security concepts in High voltage transmission EMS systems after a long career in the American power industry. He put us in touch with the utilities in Argentina where he was the prime mover in specifying the most advanced EMS system for HIDRONOR. This system required all the most advanced on-line optimisation software for a power network including security analysis, network simulator and generation management. His client (HIDRONOR) had great confidence in him and allowed him to specify the most innovative and advanced system possible. It was a long bid process and heavily fought with all the large foreign competitors involved, as we all understood that the winner would have a great advantage in future bids for EMS systems. We did win, but unfortunately it coincided with a financial crisis in Argentina and eventually the bid was cancelled.

This was the first of several times that I have won a bid only to see it eventually cancelled. Our friendship with the consultant however continued to blossom and over the years he invited me to give several technical talks at various power industry conferences around the world , which was good publicity for us. He even arranged to get some of our larger European and Japanese competitors to join his international gatherings though they were reluctant at first. One of the most interesting meetings was at Alpbach, high in the Austrian alps and another one on the island of Sardinia in Italy. I got quite friendly with some of my European competitors and even sent them our company calendar annually at Christmas. This calender was a great marketing tool in those days as our the sales department had arranged for a well known local artist to produce views of all the beautiful sites in the west island of Montreal for the calendar.

The project in Ecuador was a non-starter, but it was a great trip for me up into the Andes mountains near Quito. It is a beautiful city with wonderful views of Mounts Chimborazo and Cotopaxi in the Andes. However the airline was a bit dated flying very old and decrepit B707 aircraft. Quito is very high up in the Andes and we even had a bit of trouble acclimatising to the lack of oxygen. Similarly Bogota in Columbia was also very high up. Our expert consultant friend had given us a good reference everywhere in South America so we always had a good reception and potential clients were always interested in our technology which was of course typically North American, but as Canadians we did not have to contend with South American hang ups about the USA.

In Argentina there were many occasions when foreign exchange was in very short supply so our clients could not even visit us to check us out. So we had to finance their trips to our factory and generally host them around the Montreal area. It was a good way to get to know them better. When we visited them, they looked after us very well. Unfortunately for them their city control centers had not been updated for many decades and were very old. The major problem for them was the very high rate of inflation. On one trip to Buenos Aires the inflation was so great that I had a different price on my hotel bill every night . The advantage of course was that with our strong currency dollars we could live fantastically well even within our normal company's daily site allowance, eating only at the best restaurants. The food was fantastic, I even got fed up eating so many of their best steaks and drinking their wonderful wines. Even though the flight there was one of the longest flights possible, it was mostly in the same time zone as Montreal so there was no jet lag at all. I made several long trips around Argentina and eventually we did win the project to develop a power DMS (Distribution Management System) for the power utility (SEGBA) for the city of Buenos Aires, many years later.

Another failure was a bid for the control system for the huge Jacyreta hydroelectric power station on the Parana river between Argentina and Paraguay. We were in a reasonable position but did not handle the "financial" requirements properly. South America is home to huge hydro power resources, but there is often a shortage of money available, so in addition to supplying systems, competitors have to bring competitive financing packages too. In the end nothing much gets done. Many years later in Brazil we did pre-qualify for the system for the gigantic Itaipu dam in the south, but by then we were no longer the only supplier for large dam control systems. Our management also refused to meet some of the imposed commercial conditions of the bid so we were not considered.

In Venezuela itself we had some interesting trips to evaluate systems in the area around Maracaibo. The first time that I went there with our jovial agent, for some reason I assumed that the utility that we were visiting ie. ENELVEN, was an oil company not a power company. Fortunately we found out in good time before as my agent laughed we had asked them how many barrels did they produce!! One of the stations that they wanted to automate was a collection of diesel and gas turbine generators just laid out in an open field. In addition to SCADA, their consultant (another famous American professor and gas turbine expert) wanted them to use expert systems for gas turbine diagnostic and maintenance algorithms. So this looked like an interesting first foray into a new technology for us. As luck would have it the professor was also visiting at the time and he took one look at me and guessed we both came from the same place in India. So there was a bit of ethnic bonding and he invited me to visit him at his company in Houston Texas.

I did go to his company on a trip there while we were talking with the supplier of the regional EMS with which the GURI system would interface which was also based in Texas. At the meeting we agreed to create a joint team, with CAE leading the bid and supplying the SCADA and the professor supplying the specialised gas turbine software. I cannot remember why the offer failed, but I suppose it was money. In the power industry they quite often find that they can get by without good computer control systems just relying on simple manual electrical switch controls. It is only when they can show that they will lose money by not controlling systems properly will they get better systems. This problem does not arise for flight simulator systems, as in this case there are government regulations that insist on simulators for pilot training. So one of the key tasks that I had to undertake was to find out much more about the commercial and financial advantages that a computer control system could bring to electric power network control. As a consequence I learned a lot about power operations.

On another trip to ENELVEN, I remember I had a very short night sleep getting there, as I flew into Maicatia (Caracas) airport late the night before and had to drive the long distance up the mountains into Caracas city, as there were no hotel rooms available near the airport. Early next morning I had to do it all in reverse back to the airport to get the first flight to Maracaibo. In those days Caracas and Venezuela were very rich and all the flights and hotels were usually full. So one was always juggling trips. Every flight from Miami to Caracas was crammed with newly rich Venezuelans and their piles of newly purchased stuff from America.

After the GURI system we also bid control systems on later and smaller dams on the Caroni at Macagua and Caruachi over the years, but the client preferred to use other suppliers to avoid being tied to a single supplier. One of the interesting issues on these systems was that the bids also included all the high power protection devices in the turnkey contract and so were also outside our direct expertise.

On these marketing trips, the potential clients usually showed a great deal of interest and the presentation sessions were quite full, so I had to make a very detailed explanation about our technology and why it was unique. In those pre-Power Point days, we had to use black boards and overhead projector transparencies only. Very occasionally, when a customer insisted, we brought along a complete small VAX computer with a subset of the MMI and a demonstration version of our control system with which we could give the client a better look at our system. This of course was a big headache, as all that heavy luggage had to be packed and loaded without the help of any porters. There was also inevitably trouble with the local power supplies etc, so setting up a demonstration was not an easy job. I remember one occasion in India none of the power plugs would fit into their strange sockets, so the hotel electrician just took off the plug and actually just pushed in the stripped wires directly into the holes wedging the wires with match sticks without a thought about safety concerns. In some countries we had a lot of trouble with customs and clearances too, as they were suspicious about smuggling. Once again, India was the strictest place, the customs impounded our equipment and we had to go all over the place to get the right signatures to release our equipment in time for the demonstration. In searching for our variously impounded equipment I got a good look at the under side of airports (including the area beneath the luggage conveyors) in both Delhi and Beijing airports. Fortunately in those days there was a lot less security, so the authorities were glad to get rid of me by letting me go into the secure areas of the airports to look for my impounded stuff. As an aside it was quite surprising how haphazardly the authorities just piled up lost or impounded luggage into crowded backrooms.To find anything one had to remove all sorts of boxes including boxes of rotting food etc. In the end,our demonstrations were worth it, for nothing could wow our potential clients as much as an actual demonstration of our equipment and software. International systems marketing is not for the faint of heart!

Many years later, after we had pioneered the introduction of DMS (Distribution Management Systems) in the world, we were invited to bid for a new DMS by the city of Caracas for a system to manage the power distribution network of the whole city. This involved the use of a GIS mapping system which was overlaid on the SCADA system to allow the work crews to easily locate and repair power outages. In addition the system could automatically isolate feeder sections to allow power to be re-routed during outages. This was also a big success and there have been several follow on upgrades too

Usually when our international clients contract with us, they expect a great deal of training in the maintenance of the new system. As these systems are too complex to explain purely by giving lectures, the only really practical approach is to give "hands-on" training. This involves the client sending some of their staff to come and work with us as part of the development team in our factory. They are given some of the actual software to write and so they get a good grounding in how the system functions. As an added advantage of this approach, the client usually also implements all the user graphic display screens and inputs all the machinery and power systems data, as it is easier for them to interact with their own operations colleagues at the station. An extra benefit is that we can bid a lower price, as obviously the client does not have to pay us for their own work. Hence almost all major systems are a joint effort of supplier and customer. International clients pay a great deal of importance to this training and it is a major selection criteria. Usually we were evaluated very highly in this criteria, as we saw it as a big advantage for us too. So over the years we at CAE became great friends with our colleagues from abroad, and had them over for parties and even played soccer with them in the summers. It is one of the hidden benefits of international engineering systems projects that we end up with lots of good international friends.

8.North Americans at heart

For a long time, we at CAE were unnecessarily cautious about getting work in the USA. This is quite unusual in Canada, as Canada's largest market is the USA. I think that at first we might have thought it would be easier to win flight simulator business in the developing world or even in Europe. The idea of competing with large US defence contractors was not appealing. There was a similar mindset about competing in control systems as well. However things did change. In fact my very first line management job was to manage a new simulator department set up to design and build "repeat" flight simulators. The first such simulator was for our first American contract to TWA airlines for a B-727 aircraft. Eventually over the years we managed to win projects at the larger US airlines, which were so large that they bought multiple simulators at a time when most airlines made a big deal about buying just one simulator. American airlines were very competitive. They used to award us multiple flight simulator contracts, but they insisted that we improve our prices for each subsequent simulator. They expected us to have progressed on the learning curve and they wanted a part of that benefit too. Later on we also began winning US contracts in control systems as well as Nuclear power simulators. By now, the US market is by far the biggest market for CAE.

Repeat simulators were a management idea to try and bring down the cost and schedule for making a flight simulator. To understand why most simulators are NOT exact repeats it is necessary to understand the government regulations that are used to certify simulators. When an airline buys a pilot training simulator,they have a choice as to how exactly the simulator should mimic an airplane. There is a range of simulator types with the most accurate (and of course most expensive) simulator being the exact replica of a specific aircraft, identified by its tail number. For a long time CAE only built replica simulators leaving lesser simulators to other companies. So we had developed an engineering department and process for customising everything. This was extremely expensive both in time and money. Eventually our simulators became so good, the regulators created a category of top rated simulators that could be used by trainee pilots to convert between aircraft types by only doing simulator flying. This implied that such a simulator certified pilot could fly live on a real aircraft for the first time without any extra training. The proof that a simulator was an exact copy of the aircraft was done by the actual airline pilots doing test flights on the simulator and verifying that it flew like the real plane. It could be subjective and we had to keep tuning the various parts to make a good fit. There were of course straight forward normal engineering test routines to check out hardware and software as well, just like any normal computer based system. The data on which the simulator was built was sold to us by the aircraft manufacturer and we had developed tests to prove the simulation matched that data exactly

During my time in "repeat simulators", I noticed an interesting "experience" related issue. One of the engineering groups that I managed consisted of 5 rather mature engineers, all aged over 50 at least. This was the most productive and efficient engineering team that I have ever worked with. This team's job was to start the simulator design by producing all the marked-up engineering drawings, collecting the aircraft and avionics data, specifying the electronic interface and generally passing on all the required purchasing and manufacturing work documents. It took them barely 1 month or so, as they knew their jobs perfectly and worked as a close knit team. Any other such engineering group would have taken at least 6 months to do all that work. One of the ways they optimised their work was to use older pre-existing drawings or documents and just scribble over and mark them up. There was no time wasted on producing new documents. I wish other groups within the company would have been as practical, then we would have been able to easily cut our engineering costs dramatically. Unfortunately it was more normal in other groups to spend a lot of time on re-doing work again and again. This issue of wasted time and inefficiency is a big problem in all large companies. A major part of a manager's effort is a constant battle to get staff to stop doing unnecessary work especially the tendency to not reuse what has already been done. That is what my boss was thinking of when he invented this repeat simulator department.

The repeat simulator idea consisted of putting together the parts of the aircraft that were the same between aircraft types and copying them exactly. The parts that were different were handled by the more experienced normal simulation software engineering department. The problem with this technique was that the experienced simulation engineers refused to believe that there were enough examples of such identical systems, so in the end the idea mostly failed. At one time I did try this idea with all our B737 flight simulator contracts with some success. The parts of the simulator that were susceptible to this approach included the motion system, control loading units, the sound system, electronics, some avionics and basic computer systems. The applications software was less easy to do in this fashion. Flight simulators were the most complex systems done at CAE as they were built with a tight system combination of mechanical, electronic and software systems. One of the key elements was the very expensive visual system which was a separate system usually bought from a specialist company. For the rest we manufactured everything except the computers. Our motion systems were considered the best in the world. This is very important as the motion gives an exact feeling of the acceleration and attitude to pilots. For military fighter aircraft we also supplied a special flight suit that gave the exact feeling of G forces in sharp turns. A side benefit to me was that the evaluation test pilots who reported to me, gave me the chance to try and learn how to fly some of the aircraft simulators on the test site.

One interesting marketing item that is worth noting is who the real selection authority is when buying a system. In control systems and nuclear simulators the key selection people are engineers, so it is important to try and impress them with technical innovation. In flight simulators the key staff are all pilots, so they are more impressed by training aids and simulator fidelity as measured by flying the device. However even with flight simulators there were a few outlier buyers such as KLM airlines who paid a lot of attention to technical details and so were a good candidate for the first development simulators. In the end technical virtuosity seemed to matter to many of our clients. In fact KLM airlines gave us at CAE and me in particular, a lot of trouble persuading them to accept the simulator as being operational. They were very hard nosed businessmen and always drove a hard bargain. However once satisfied they always gave us a superb reference. So it was always worthwhile to have other such tough clients as initial customers, as it paid off in the long run. One consequence of the usual contractual terms was that failure to deliver on time was to get a price penalty. Almost every single customer in all our businesses insisted on such a contractual clause. As we were generally delivering brand new technology, we were occasionally late, as it is very difficult to predict strange technical issues for the first time. So at final contract negotiation, just before delivery, there was a long negotiation session trying to get to a compromise, whereby we always delivered more than the minimum in lieu of delay. Our contracts manager even invented a negotiating ploy he called "excusable delays" to get a better deal. It was all good fun, and our clients realised that they always got the best system possible at a very good price. When one considers that our average system was used for at least 20 years at a stretch, it was always worth waiting for the best technology. In fact some of our control system RTUs have run for well over 30 years.

Once we won the first American commercial flight simulator, we suddenly began to grow very fast and soon became the very largest flight simulator company in the world, which CAE probably still is. Our company president had arbitrarily decided that we would significantly reduce the schedule to deliver simulators well down from the usual 2 years plus. At the time I did not think this would work, but it did and it led to many clients only wanting a CAE simulator. This also gave us the confidence to start bidding for other systems in the competitive American market, both for Nuclear power simulators and for computer control systems.

About this time Ontario Hydro (OH) had started evaluating the Pickering A nuclear operator simulator that we had built for them. As they were very happy with the advantages they got from the training, they decided to buy a simulator for each nuclear reactor power station that they owned. Suddenly this created a huge amount of work for us to build 4 very large simulators. The company had to create a special department just for nuclear simulator modelling application software. The other computer software, electronics and mechanical engineering was left to the conventional departments. A big difference between flight simulator software and power simulator software was the enormous size of the power simulators which had to simulate many times the total number of devices. A power station is full of dozens of pumps, pipes, motors, boilers,heat exchangers in addition to high voltage electrical distribution circuits,the complex feed water system, large steam turbines and electric generator systems. To model all this needed an army of modellers. The Instructor facilities on a power simulator had to correspondingly also have to use several CRTs in order to show the training staff a graphic image of all the panels locally. This was so that the trainers could observe what the trainees were doing without disturbing them working at the actual panels. One consequence of this was that we invented what was essentially a very realistic set of virtual computer generated "glass" panels. A few clients even used these CRT based glass panels to avoid the cost of building actual hardware panels. Of course the highest profile simulation model was of the nuclear reactor itself. The physics of nuclear reaction was modelled by dividing the device core into a large number of sections and calculating out the interlaced effects. For the CANDU reactors we also had to model the moderator system as a whole, but it was not necessary to model the fueling machine though it was mimicked on the hardware panels.

To build an exact copy of the station control panels and instruments is a very difficult though tedious task. The reason was that when a station is built there is no standard production type repeatable control panel design. So the hardware engineers had to visit the site and take photographs of the panels and instruments. Some of the instruments could still be purchased, but many were no longer available and we had to actually manufacture them in our factory from photographs and measurements. This is a huge job as these control panels are enormous, often taking up an entire large room a couple of times the size of a typical tennis court..

The reason why nuclear plants have such large though antiquated control panels is probably mostly historic. Usually power transmission control centers are mostly computer controlled and use graphic displays with virtually no traditional instruments. Fossil stations are usually a mix, with some computer control, but still do include a few traditional panels. Very modern station designs often are all computer control but there are few such stations that have actually been built. Canadian CANDU reactors had a mix of computer control and some panels. I think the reason why US stations were more cautious could also have been the fact that they were developed from military reactors where submarines did not have the environment for computers. With hind sight I think there is a good Human Factors reason to use panels. In an emergency, panels are easier to read and understand. Many years later I was invited several times to visit the US nuclear regulatory authority in Washington. This was because they wanted to understand the benefits of computers from the practical experience of us Canadians. In the end they remained conservative, though they kept praising our forward thinking. The most advanced nuclear controls that we ever did, were for the Darlington station in Canada where the panels had been minimised and many functions moved to the station computers. I was never personally involved in CAE's forays into power simulators in Europe, but I have seen some designs that use a great deal of computer based data-logging.

The other item that we pioneered was integrating the actual station computer control system into the design, so that we could use the actual control and monitoring software. Especially for CANDU reactor stations, this computer control system is closely linked with operation of the station. It has to be an integral part of the simulator, otherwise the station as a whole will not behave realistically. This was called stimulation as opposed to simulation. This approach is fairly common nowadays as most aircraft also use flight management computers and EFIS engine control computers , but it was very unique then. In fact on our very first power simulator, there was even some doubt if such an approach would work. So the client, Ontario Hydro, gave us a let out clause in the contract. Fortunately it did work as we found some spare areas of memory within the heavily loaded control computer that we could modify to put in Instructor facility break points. If a control computer has to be used in this way, one has to get into its details so that the control software can be restarted at different points as required by an instructor when he puts in malfunctions to test a trainee. As the control computers were quite old in most power simulators and had little spare memory or capacity, we had to spend a lot of time understanding this control software and modifying it carefully to put in break points for the instructor. Later on when we began building nuclear simulators in the USA, we found the station data loggers were so difficult to evaluate, we usually ended up running a software emulation of the entire computer system in a more modern computer.

CAE was quite carefree in redesigning items to make an exact fit. So even though we already had standard Digital input and output PCBs, we designed new larger PCBs specially for the power simulators as they used very large numbers of inputs and outputs on the panels. In that way we could fit all our interface electronics at the back of the actual control panels in DATAPATH C chassis, strung along a DACBUS. Then to cut costs for our flight simulators, we began designing electronics that actually fitted on the base of the moving simulator cockpit, thus avoiding long cables drawn over to outside fixed cabinets. By putting intelligence locally in each chassis, we even had on board interfaces doing much of the work for aircraft instruments and circuit breakers locally. The use in modern aircraft of flight computers also led us to new designs based on stimulation of those computers. In many such cases the equipment suppliers kept their code secret but made special arrangements to allow instructor facility inputs to be made. Alternately one could always emulate the device as well in the main simulation computers.

In the USA power simulation was a rarity as there was no regulation requiring it. Then an accident occurred at the Three Mile Island nuclear power plant and all hell broke loose! As a consequence, the US nuclear regulators mandated that every nuclear station in the US had to have a full replica simulator. Suddenly a whole new market opened up for us in the USA. At first it was difficult to persuade the American nuclear station owners that our experience with CANDU reactor simulators was valid for the American PWR/BWR reactors, but eventually we won a reasonable share of the market. In the end we probably became the largest supplier as the other American suppliers could not make enough money and mostly bowed out. This business is still going strong with many simulators supplied to China, and Europe, in addition to CANDU simulators in Korea , Quebec and New Brunswick. I travelled to some interesting sites in the US far west such as WPPS in Washington state on the Columbia river where there was a huge nuclear industry built during the atom bomb project. The Americans had built large farms in the area, especially for the grapes used in wine making. I have never seen such enormous fields irrigated with water from the river over a series of giant automatic sprinkler systems. The Americans have created new fertile lands in what was previously mostly desert. As a contrast I also had the chance to visit stations in the East such as at Indian Point on the Hudson river near West Point where the local culture and thick forest environment was quite different.

Modern replica simulators (both aircraft and nuclear power station) are now so accurate, they are often used by the regulatory and safety authorities to verify conditions. This is also useful after an accident as one can actually observe how the various systems did perform, provided there is enough data collected during the accident. Most advanced industries have data recording devices that continuously collect critical data while under operations. This data can then be fed retroactively, when checking out a possible failure scenario, using the Instructor Facility (IF) programs to inject the new data and start from a test situation. All our IF stations allowed this type of analysis and recording. In addition even all our own design computer control systems did collect such data continuously for after the fact analysis.

One of the innovative developments from the power simulator department was the development of the first graphical programming language ROSE (Real-Time Object Software Environment). This allowed simulation software to be built by just using pre-designed graphic icons similar to the process design sheet logic. Each specialised icon was a software object that called up the complete simulation behaviour of the object eg a pump, or a generator etc. Object oriented programming languages are fairly common now but then it was quite unique and gave us a big advantage. We even used this language as a design language when developing naval submarine machinery control algorithms and other designs. Interestingly this language was developed quietly as a sideline without any actual executive permission. It goes to show how tolerant our bosses were in giving us engineers lots of room to innovate.

Another advantage that we had was our very detailed and clear documentation and design approach. The Quality standards that we used were the stringent standards for nuclear control systems in Canada. The reason why we followed this approach was because our first customer was Ontario Hydro who were an engineering organisation with very stringent QA requirements. In the USA, our American competitors were only used to simple commercial standards as US stations at that time built nuclear plants as if they were fossil plants. This level of design management, impressed our American customers, but I didn't realise this until I was told it at a nuclear conference in Tennessee. This senior engineer told me that it was this rigorous approach that gave us the reputation for getting the best engineering and technical evaluation. Even at CAE, to keep costs down on flight simulators we just used commercial QA standards, but because of our history on nuclear control our control departments used the significantly more stringent standards.

This QA approach paid off many years later in quite another field, that of military support for the Canadian air force fighter aircraft. At that time there was a competition to set up a support facility for the CF-18 fighter aircraft in Canada. Our team consisted of the prime contractor who would maintain the airframe,engines and aero-instruments and we would maintain the electronics and software. The Canadian air force officers, who were familiar with our commercial flight simulators were suspicious that CAE was not up to the stringent QA requirements of the project and wanted to re-evaluate our bid. However we quickly modified our bid claiming that this work would be done by CAE's control department using our nuclear standards which were even higher than the customers. So that is how I got involved in CF-18 aircraft instrument software.

We could not replicate our success in nuclear power simulators in the field of fossil station simulators, as that part of the industry was very price conscious and there were no regulations that mandated simulators. During my trips to South East Asia I did try to interest some potential clients, but nothing came of it. There was only one big project win for a pair of fossil simulators for Ireland. However by studying the fossil plants we did eventually learn enough to use similar simulation for our marine control machinery simulators

The first computer power control systems that we won in the USA were also in the far west. There were at least 4 other large experienced American suppliers, so we had a hard time breaking into systems in the east. Somehow it was easier out west. This first system was for an unusual system that managed the entire Columbia river basin, which consisted of several hydro-electric plants on the river. This system was responsible for coordinating all these plants on the mid-Columbia section. This control center measured and allocated flow to each station on the river taking into account the legal requirements of water flow for irrigation,fish ladders and tourism in addition to power output. It was all very complicated and required a new software application called Energy scheduling and accounting with a large database to handle over several thousand accounts. To handle this we invented a new type of data base system as at that time it was felt that commercial database systems would not be able to do this job fast enough. Usually contracts went to the low bidder, and as the Canadian dollar was low then compared to the US dollar it was an advantage for us.

A little later I was on a power systems course at University of California in Berkley and had just entered the elevator when I bumped into a figure from the past. He was one of the GURI consultants, who had been impressed by our GURI system. He was in charge of the controls at the Grand Coulee power dam and mentioned that they were already in the middle of bids for a new system replacement. He was so enthusiastic about our technology that in spite of our caution we did bid on that system though expecting that we would lose to an American company. One should remember that the Grand Coulee dam was the largest in America at over 6000MW, so it was a prestigious project. We won on both technical and price comparisons, though we had to wait as other American companies made all sorts of objections to awarding such a high profile contract to a foreign company.

I was very impressed during my visits to Grand Coulee dam. My friend showed me every detail all over the power house and its associated irrigation works. What most impressed me was that the power house was kept extremely clean, almost spotless compared to the messier dams in China and South America. The system we were replacing had one strange item. The mimic was just a set of several CRT monitors hung up from the ceiling. At first they wanted something similar from us, but in the end they came around to the usual tiled mimics that everyone else used. I have only seen one other such CRT-TV based mimic at the Philadelphia Electric company control center, when we visited them for their control center upgrade. Over the years I made several visits out west for other systems. The region had huge public utilities such as Bonneville power and WAPA, unlike the East where utilities were smaller private companies. The US federal government was the only organisation that was willing to develop this area. One peculiarity of this was that our clients were government employees who had to follow very strict regulations about accepting gifts. So during meetings they only agreed to accept the usual coffee hospitality if they themselves paid !

Grand Coulee had a number of existing generation management software applications and our task was to just copy them onto the new technology. For the rest there was the usual SCADA system software. One unusual item was the very large scanned database. There were 40 RTUs, as there were a very large number of generators and a total of over 39,000 scanned data points. This made the system the largest electronic hardware interface that we at CAE had ever built for a single customer. In those days the CAE manufacturing department did not pay adequate attention to control systems as the bulk of their work was for flight simulators. It was only when I told them that Grand Coulee alone was the size of 10 or 12 separate flight simulators that they had a shock and re-planned their work.

In order to work with the power utilities,it was necessary to have detailed knowledge of high power electricity. The North American and British fashion in those days when studying electrical engineering, was for everyone to chose electronics rather than high power. So we had a great deal of trouble finding engineers who specialised in high power electricity. In the end one of my colleagues who was also a professor at Mc'gill university, found a researcher in the subject. We hired him and for years he was the only power expert that we had. Interestingly enough, high power electricity had become important again, as utilities were using modern computer techniques to control power networks. As a student, I remember all of us were very scathing about the few who did take the high power option,as we considered it old fashioned nineteenth century technology,while electronics was considered the most modern technology. One never knows what is going to be useful in the future!

At around this time the available computer technology changed. Among the new elements that we introduced were the world's first use of the dual Ethernet LANs to connect all the elements in the master station control room. This network approach became the norm on all our systems including the various simulators. For quite a long time we were unique in this approach, but now all the suppliers use LAN technology. We also upgraded both the front end system and the HMI graphic system by using CAE designed microprocessor units. At this time Intel had just come out with new chips that integrated full pixel graphics with a cache for pre-coded characters or symbols. This allowed us to create a single graphic card that could multiplex character type graphics melded with full vector graphics. That way we could develop more complex full graphic displays than on the standard Aydin, but running at the same speed. These displays had the speed of raster graphics and the detail of vector graphics. The CAE front end was based on our new DMC Intel based processor, which included 16 serial channels on a single board together with the required memory for the front-end processing and for computer networking to external computer systems. On very large systems, we could just put in more such front end boards to create a very powerful front end communications multi-processing system. This same technology was also introduced into the RTUs, so we no longer needed the stand-alone Intel 8086 of GURI days. Eventually this same in-house approach was used all over the company in marine control and simulators. Even the large VAX main computers were replaced with distributed micro-VAX computers as several micro- Vax computers were cheaper than a single large VAX. We found out that our master station software was well designed and modular enough to be easily re-allocated across several distributed computers without any loss of speed or function.

At this time we had a spurt of new orders based on our unique technology,as our competitors took a very long time to revamp their designs. Even our use of LAN technology could not be copied for a long time, because the competitors could only use a single LAN. We were unique in using dual LANs and so had higher availability numbers. At the same time I began to suspect that it was time to change our communications software technology by beginning to move away from proprietary protocols to OPEN standard telecommunications protocols. In the past, all SCADA type systems did all their networking and communications using proprietary company specific protocols. The initial arguments in favour of this approach was that custom protocols were required because control networks had to be fast and economical. As the IC industry had started developing communication standard hardware, it was time to gradually redo our communication and front end software to the 7 level ISO international standard. This was another advantage our systems had over our competitor's proprietary protocols. I was too optimistic in my estimates as it turned out to be a very big task, but eventually it paid off as the entire industry moved to the international telecommunication standards, so that SCADA could piggy back smoothly onto public networks too.

Another advantage that we had at CAE was our military /nuclear background, which used, as previously mentioned, very stringent QA unlike the more casual QA at our competitors. We even did all our theoretical reliability calculations according to the latest algorithms and exact IC chip data. This was possible as CAE had a complete department that specialised in just these R&M (Reliability and Maintainability) analyses. There is a military/space-flight specification which makes it necessary to do this detailed calculation when quoting R&M figures. In this calculation, the data used are the actual R&M values that have been measured for virtually every component used, including every IC on every PCB. This is a major task and this department actually monitored all our delivered equipment and used the actual achieved reliability to update their data base of values.When evaluating systems it is important for a serious customer to see such accurate data, as it gives a better expectation of a more robust design. However in the industrial world most suppliers do not go into such detail.

During this period, buoyed by our victory we visited many other potential clients out west in California (San Francisco, San Diego, Los Angeles), WAPA (Western Area Power Authority), Idaho power,Seattle city power, Portland and so on. One peculiarity that I remember vividly was the arrangement of the control room at Seattle city power. That room was so small they had the mimic board behind the operators as that was the only place where it would have fitted! It reminded me that as most US power utilities were private companies fixated on profits, they spent the least amount possible for automation. That resulted in absurdities like this mimic. The state owned utilities such as Bonneville power or USBR (US Bureau Of Reclamation) or TVA or WAPA were a bit more like our government owned utilities in Canada.

One of the sales strategies that we followed was to attend all the annual American IEEE conferences on power automation. I had to give presentations on our systems and we usually built a demonstration system for visitors to look at. I generally found these events quite boring, but the sales department insisted on going. I doubt that these meetings pay off on project wins. They do however give one a chance to get market intelligence on upcoming bids and on what our competitors were doing. In any case, I preferred to visit individual potential clients for one on one meetings where we could discuss their very specific issues. Some of these IEEE conferences were also held in Europe and South America. I even had an embarrassing experience during one of these conferences. On one nuclear power meeting in New Orleans I took some potential clients for dinner at a very expensive Creole restaurant in the old quarter. Unfortunately I had not looked carefully at the entrance notice which stated that one could only pay cash, so when my American express card was rejected, as I did not have enough cash on hand, I was forced to ask my guests to help out! One would have thought that I would have learned a lesson from that, but I had a similar problem in Sweden once when the hotel refused American express cards but fortunately it accepted my personal Visa card! Such are the occasional hiccups of international travel.

Then we had a lucky break. For several years we had been trying to win a real full EMS system, as that was a bigger market, but kept failing, as our competitors had much more experience, especially in the crucial area of power transmission security software applications. PSEG (Public Service Electric and Gas) of New Jersey was one of the largest utilities in the USA and the pioneer of EMS systems at their control center in Newark NJ. They were an early adopter client and hugely knowledgeable. So they came out with the specification for the most advanced possible EMS design for that era. Their existing supplier was also the biggest in the business and normally they would have gone back to them, but this time they decided to go for a competitive bid. After they checked us out they became quite enthusiastic about us, as we usually were the most advanced in general systems software, computers and electronics. Their new system was to include all the then emerging new technologies such as a distributed LAN master station, full graphics CRT HMI, OPEN standard communications,RDBMS (Relational Data Base Management System)and the fastest possible computers. The software had to include editors, databases and graphic displays of a type that were unknown then, but are fairly usual now. One could think of this systems software being like the modern WINTEL PC HMI and applications of today, but developed at least 15 years before on much slower computers. For example we invented various types of "calculated" points made up from various scanned or other calculated points, giving us a primitive version of a modern spreadsheet. Our editors started to become WYSIWYG (What you see is what you get) type of editors rather than old fashioned line editors. Our homemade math and graphics packages were similar to modern software packages. We even had graphic MMI software that could put up multiple windows on screens just like modern browsers. As I have said earlier "there is very little truly new under our sun", especially as this was only in the early eighties. So we bid the very largest VAX computers of that time and one of the earliest available full graphic CRT workstations from a new start-up company called IDT. To offset our weakness in power applications we offered to convert and re-port their existing power applications suite onto our new system without the software slowing anything down. It turned out that they were delighted by this offer as they were used to their own applications and had spent years getting them optimised for their own power network.This also turned out to be a lucky strategy as our competitors refused to take the risk of offering all that new technology and development, preferring to bid their standard systems.

So that was our first real full EMS system and once again we became one of the industry leaders. This client has been good to us and the relationship has lasted, with us winning several upgrades over the years. One of the problems that became very difficult to resolve was the IDT workstation (which was unreliable and that company eventually disappeared), We had to replace them with Silicon Graphics workstations. By then we had a lot of experience with Silicon Graphics as we used them for the Instructor facility on our simulators. As the system became bigger, we had to update to larger and faster VAX computers. Another advantage was that their power engineers taught us a lot about how one manages power security applications. Their expert was a very kindly gentleman who took a lot of time at power industry conferences praising our software. At parties for the team he was also very proficient in playing the piano.

Once again this episode showed me the importance of listening carefully to a client, rather than thinking we were the only expert. As a consequence of episodes like this, some of our larger competitors began to leave the business, making room for newer more responsive competitors. Sooner or later all these bigger companies were sold off, some several times. It became difficult to keep track of who owned which company. The Europeans who never won any large American system, eventually had to settle for buying American companies to get into this market, which was of course the largest in the world.

Finally another bit of luck. The Union Electric (UE) company of St. Louis was looking for a new system to replace an old SCADA system which had been supplied by one of our competitors. They appeared to be happy with this supplier and our expectation was that it would be pointless for us to waste our money on the bid. However our sales manager had discovered that the client was actually looking for a new unusual type of system involving very high speed graphics. As we already had a very skilled software team working on complex graphics for our flight simulator Instructor Facility group, I agreed to check out the opportunity, even if we were unlikely to win. So off we went, but got stranded at night in Chicago's O'Hare airport as a storm shut down the airport. We finally left at around 3AM in the morning for St. Louis and we were not too positive about our ability to stay awake and make a good impression at the meeting which was to be at 8 AM! I am glad we went, as it turned out the client was going to develop the world's first ever modern full DMS (Distribution Management System). Let me explain.

Even though SCADA and EMS systems had been used for managing generation and transmission systems for many years, distribution utilities had few incentives to bring in real time computer automation to manage their systems. The first problem was of course money. Distribution companies are the poor relatives of the power industry. They can rarely justify spending money on computers, and are content to work with paper based administration. The distribution power networks have virtually no automatic remote control on their switches, or even any remote instrumentation for electric signals, except at very major substations. Faults are fixed by sending work crews to drive around to look for the failed area and manually re-route power. The main way they find out that there is a problem is when a customer rings up to complain. They have offline customer information systems for financial purposes, but this information is not connected to the control center facilities.

The other major draw back to automation is the absolutely enormous size of the potential database and the huge number of signals that would have to be monitored. This is several orders of magnitude larger than for a transmission system. In fact earlier computers would not have been able to manage such huge databases in real time. Modern computers made this finally possible. The graphics industry had just invented Computer Aided Graphic and mapping systems (GIS) that allowed an entire city to computerise all its facilities onto a graphics data base and map, eg. the power network, the streets, the water supply etc. As the city of St Louis had just bought such a system and put all the city's facilities onto computer maps, the power engineers thought that they had the possibility to use this existing computer data to invent a new type of DMS system.

Our lucky break was that they awarded us the contract as their existing supplier was very negative about it and refused to believe that it was even technically possible to integrate detailed graphic maps onto a modern SCADA system. The problem that everyone believed, was that a graphics map overlaid SCADA type picture would take so long to come up on the CRT screen, no operator would like to use such a system. They would probably go back to using paper maps, with pencil markings indicating work orders. Older distribution control rooms just consisted of rows of telephones and huge racks with large paper city-maps on which the control operators did all their work. UE wanted to move all such management and control activities to a modern SCADA and full graphics integrated DMS system. They even wanted to put all their work order and outage management activities onto the same system, with radio based laptops in each work truck, so that crews could be directed to work sites automatically. They had also increased the amount of remote automation in some substations, though much of the database was still updated manually at the center by the operators depending on what the field crews had radioed in. Even though this was a very difficult and quite unique system, when we finished, CAE was the only possible supplier for this type of system, and it led directly to contracts for many of the world's largest DMS systems.

The design consisted of the usual SCADA master with dual VAX computers, CAE front end communication systems, inter LAN networks to other company computers, dual

SCADA LANS and a large number of Silicon Graphic work stations for the operators. The client supplied the radio systems. We had to develop some distribution specific power applications in addition to the trouble call and outage management applications. The highest profile software was of course the detailed and very fast integrated graphic systems that worked at the same speed of normal SCADA pictures when calling up the full graphic map pictures. Our new graphics software allowed panning and zooming across multiple CRTs in much the same way as modern Google Maps does today

Such a modern DMS now has the following software applications:
Switching Management, Connectivity analysis,Power flow analysis,Demand estimation, Distribution optimisation,Fault management, Outage management,Fault level analysis, and Volt-VAR control. Some of these programs are variations of the kind of algorithms developed to manage HV transmission networks which I have described elsewhere. There is a lot of guesswork and customizing of the software and data to make a reasonable fit with the actual network situation. The outage and fault management programs are just computer automation of the administrative procedures used when directing field work crews who fix the problems on the power networks. Since fully hardwired instrumenting of the power network is very expensive, it pays to develop software that estimates the missing instrumentation in order to give a more accurate description of the system to the control operators.

In fact the data and operator work load can be so great, one of the control operators on our next system for Boston Edison, told me that during a major emergency, they actually go back to using manual telephone, paper and pen techniques to control multiple field work crews in parallel. In such an enormous emergency, the computer system was used mostly to register events for later analysis and to finally re-institute the status of the power database correctly. It highlights an interesting problem, eventually computer software is essentially a sequential activity, however fast it does its work. Managing an emergency can be more chaotic with several things going on in parallel. The human operators have to use their ingenuity and skill to do stuff in parallel. This Boston Edison system was an enormous system for the whole of the city of Boston. and the client was very helpful in aiding us by showing off their system. As it was relatively close to Montreal, I drove potential clients from Australia, over the White mountains of New Hampshire to see the system. We got a very good reference from them, and they have returned to us for upgrades.

9. The great wall of china

Another chance telephone call resulted in opening up a huge market for CAE in China. At the time China was just coming out of the era when China was essentially closed to the world. The opening to the west had just started and as part of this, the Chinese had decided to invite the General Electric company to visit China and give a series of talks on their new hydro turbine technology. The GE hydro turbine department was based in Montreal quite close to our factory, so they called us to see if we could support their talks with discussions about our hydro control equipment, as they themselves did not have that technology. Initially I was dubious about the value of going there as I did not expect there would have been any payoff in terms of new business, but in the end the chance of sight-seeing in China was too tempting. Just as well, as eventually the visit directly led to CAE's first contract over there and over the years China became a huge market for CAE, especially as we had the first foreign supplier advantage. This was even true for our flight simulators as our then sales manager for China began to learn how to do business in China and was able to take over selling flight simulators as well as control systems. In the end with our familiarity with the Chinese market we even got a good foothold in their nuclear power simulator market too. It was a classic tale of how a business can grow from nothing in a new and unknown international market, simply by having the courage to give it a real try.

The visit was organized by YVPO (Yangtze Valley Planning Office). This was the regional authority that had the mandate to develop and manage the whole Yangtze river basin. The target project was the Three Gorges hydro-electric facility that was under review. If it was implemented, the project would have been the largest hydro scheme in the world. As a consequence of our experience on GURI,Grand Coulee and James Bay we were the only controls supplier, at that time, who could have met their criterion for selection. However as the Chinese were not aware about these external developments, they were surprised and eventually impressed with our technology. Our trip by air was through Tokyo where I nearly missed the airplane connection to Beijing. In those days there were so few international flights into China, that if I had missed that flight I would have been unable to fly there for several months.

Landing in Beijing was very interesting. The old airport was very quiet, as there was not much traffic. On the drive from the airport to the hotel, I saw how well organised the Chinese were. Even though the road was probably several miles long there were small plant pots with flowers lining the entire road, with no gaps, all the way into town. It seemed to indicate a mind boggling governing ability to organise that. There were only a couple of hotels for foreigners in those days. As China was just opening up, the Chinese were only just learning the required features of good customer service. For instance when we went to the hotel restaurant for a snack, the waitresses ignored us and kept on chatting among themselves. They were generally quite rude (as good communists were not expected to serve anyone). Eventually on future trips I began to see an improvement in service. This was brought about by the Chinese hoteliers themselves importing local staff from Hong Kong who brought their superior customer friendly traits with them. In those days I could easily spot a Hong Kong waitress from a local Beijing girl in just a second. The local GE office organised our visit with the YVPO.

Accommodation for foreigners was kept short but GE were a favoured supplier so they were allowed to set up some offices in the foreigner's building. Their manager had to live in Hong Kong and only came down to Beijing for short spells. He was very good to us and gave us the full tour of Beijing and the great wall. In the end I made so many trips to China that I got quite good at finding my way around Beijing and seeing all the usual tourist places including the summer palace and zoo. On each trip there were more and more hotels available, including some of the most luxurious that I have ever stayed in.

We were taken on a trip around the towns of the mid Yangtze basin starting with Wuhan, where the main discussions were held. Each technology ie. Generators, turbines and controls were discussed in different rooms with Chinese engineers doing the interpreting. As computer control systems were quite new in China, they had a great deal of interest in my talks. They were fascinated by my descriptions of the GURI system, to the extent that they insisted that I make an instant design of a proposed computer control system for the proposed three gorges power network. They only gave me one evening to do the design, which I did on the large sheets of paper that they gave me in my hotel room. The next day I described the design in detail. They were interested enough that they took it and put it on the walls of their offices at the YVPO and eventually years later they used it as the basis of the actual control systems design.

Basically I came up with the idea of creating a cascade control center based at 3 Gorges, in addition to a station control center. Eventually the Yangtze river including its tributaries is going to have a whole slew of hydroelectric run- of- river dams. So really it becomes very important to manage the water flow so that it is optimally allocated to various facilities before it all flows down to the sea. The river is used for water transport, irrigation and power. As the weather can vary enormously, flood control is a major concern. So for accurate river management, the central computer had to be able to make reasonable forecasts and using good simulation software allow the staff to plan secure control strategies. This software had to use inputs from all sorts of weather/rain gauge stations across the river valley while making allowances for the condition of the soil etc. The system would have required a very extensive communication network to connect all the different computer control centers together with all the outstations.

This internetworking of many diverse systems is getting quite common now. At that time it looked very new and exciting. This was before there was enough of a commonly available extensive digital communication infrastructure anywhere. So I expected that one would have to actually invent all the appropriate protocols, hardware and networks. Only the DEC company had invented enough software for such a network called DECNET. So our use of VAX computers came with the advantage of both DECNET and Ethernet LANs. Another missing element was a reasonably inexpensive relational Database system. I assumed one could just use something proprietary like our own SCADA database. In today's systems many of these problems no longer exist as all such software is easily available. Many of the systems that we designed in later years were actually large networks of different era computers exchanging all sorts of control and background data through specialised servers and LANs. So in modern real time computer control systems there can be an enormous hidden computer systems and communications network overlaid on the actual physical systems that is being managed.

Over the years I made several technical talks to Chinese engineers at various seminars and meetings. In some of the talks there would be around 80 to 100 participants and they all seemed to have lots of questions and used to write down feverishly what I was saying (or what they thought I was saying). When the interpreter was a fellow engineer it was quite easy, but at some institutes the interpreters were fine arts graduates and it was very difficult for them to translate our technical words. At one of the more amusing meetings in Tianjin, I made a single sentence description of some technical feature, and was horrified to see the interpreter take nearly 10 minutes of elaborate sounding Chinese words to translate what I had said. I expect either no one understood what I was saying, or most of the engineers understood enough English to get the gist of what I was saying. Of course every meeting was followed by a huge banquet with lots of polite conversation and plenty of toasts made with fiery mao-tai liquor.

The YVPO engineers took us by train along the Yangtze river to the site of the Gezhouba dam at Yichang, that was being built as the first phase of the entire three gorges project. It was an enormous run-of-river plant built by a huge team of several thousand constructors. One of the power plants was already working while the second plant was being built. It had a simple Chinese computer SCADA system. As it was difficult to import foreign computers and equipment they had designed it using off the shelf commercial equipment such as ordinary TV monitors in place of CRT displays and generators. Even the system clock was a normal mechanical clock with an innovative interface to the system. The RTUs were made by a local design institute the NARI (Nanjing Automation Research Institute).

As a treat we were taken by boat up the Yangtze river to the construction site where the 3 Gorges dam was going to be built. It is a beautiful ride with steep canyon walls on either side. One of the construction engineers gave me a sample granite core piece from the dam site, which I still have and use occasionally as a paper weight. The river has such a fast flow, it took all morning to work our way up the river, but we returned down river in barely a couple of hours. Even though I love Chinese food, I had previously never bothered with chopsticks. On this trip for many weeks there was no alternative to learning to use chopsticks. We Canadians used to compete at lunch as to who had the most skill. The most difficult task was to be able to pickup a single peanut lengthwise without dropping it.

I learned how engineering work was allocated in China. As China was a communist country, they did not have commercial companies of the kind we had in the west. So engineering work was allocated to different technical institutes in different parts of China. Each institute was allocated a different technology or a different region of China. It was all supposed to be a quite rational way to allocate work. However behind the polite facades, the various institutes were deadly rivals. Each one wanted to team up exclusively with prominent western suppliers such as CAE. Of course at the time I was unaware of these machinations but got quite familiar with the Chinese system over the many years that I became familiar with China. In fact a major clue to business success in China was to team with the favoured Institute for each project. Choosing the wrong institute led to automatically losing the project. Teaming with the right Institute ensured success even over other internal Chinese bickering. Some years later while competing on an EMS system for Hebei province, we had done the bid and were about to be rejected politely by the evaluating technical agency. To let us down softly and politely, that agency invited us to a banquet to give us the news that we had lost. However they had not organised things properly with the actual project customer, who cancelled that selection and just chose us over the selection agency's choice . So we got the contract for the Hebei EMS, in spite of the influence of the selection agency, simply because we had the support of the Hebei power utility with whom we had done our homework!

Of course this is getting ahead of myself. First we had to win the most prestigious project at the time in China for the Gezhouba control system, which was then the largest hydro-electric facility in China. This started in an oblique fashion as I was on my way out at the airport in Wuhan when the chief engineer of the YVPO actually held up the flight so he could make sure to give me a technical description of the required system for Gezhouba. Foolishly at the time I was unaware how important such a gesture was. I almost turned it down as being too small for our company, not realising the importance of such a gesture in the Chinese political environment. The Chinese were not to be put off by a foolish foreigner like me and gradually worked their way in. Along the way they visited me in Montreal. and independently checked out our references at GURI and James Bay. This included a trip with me to the far north of Quebec to see the James Bay system. Later on I was told by one of our Chinese contacts, that they somehow had a great deal of confidence in me as an engineer and had therefore selected us as the preferred foreign supplier to build their Gezhouba control system. This contact happened to be a close relative of the first president of China and he was also a professor in BC (British Columbia). He was discovered by our executive in charge of fostering international agents. As CAE did a lot of international work they had allocated a senior executive whose main job was to find influential potential agents in every major region of the world. As a consequence this executive had a most interesting job with lots of connections around the world. To get help he was delighted to let us also take a role in selecting the agents,so I fell into the habit of trying to build up contacts all over the world as well, but especially in Asia. This Chinese gentleman was most helpful in our first bid. When we were negotiating in Beijing, the local Chinese purchasing agent kept referring to him and implying that they were doing us a favour just because he was so important to China. I don't know how true that was but it did make a good negotiating ploy.

The Chinese pay a great deal of importance to personal relationships rather than just relying on written contracts. During the bid evaluation meetings in Beijing on several occasions the purchasing organisation kept pressuring us to lower our price, to the point where we wanted to throw in the towel and go back to Canada. The Chinese project people would not allow us to falter and every evening came to our hotel and forced us to carry on regardless. That is how we finally won the first CAE project in China over much larger better connected foreign companies, simply because the local project people played the internal Chinese political game better than the competition.

One of the elements in the selection process was for us to team up with NARI (Nanjing Automation Research Institute) based in Nanjing. We were thoroughly clueless about the process. One day we visited Nanjing for another reason, thinking we would just pop in to say "hello". We did not realise that we were being led there by our potential customer who made sure we did exactly what they wanted. In those days whenever I visited China, without telling them that I was coming, somehow they always knew that I was going to be there and were waiting for me at the hotel. I ended up having the highest regard for Chinese organisations. While in Nanjing we missed the plane to Beijing, to catch our flight to Canada, and that was a problem as we would not have been able to get another flight for a month. They solved the problem by quickly rushing us onto a train to Shanghai, and we made it to our connecting flight to Canada just as they were closing the aircraft doors. I have had several close calls in my career taking flights all over the world, but that was the closest that I got to nearly missing a flight. In those days China had a system whereby almost all flights started from Beijing only. So even if we wanted to fly to somewhere in the south from another city in the south, we had to go back to Beijing first. If we missed a flight for example due to fog, then we lost our turn on the flight and had to go to the back of the queue and wait for several days. I can remember being stranded in Guilin for some days because of this. Eventually a few years later it became possible to fly into south China directly from Hong Kong and that was a relief. Another quirk of the early days was the issue of taxi cabs. In those days taxi cabs belonged to the state and each driver was given a fixed amount that he had to bring in every month. After he made his quota, he could give up as he could not keep any part of what he made above his quota. So naturally all the drivers stopped after a couple of weeks into the month and there were very few or no cabs available towards the end of a month. Of course we couldn't understand why we could never get a cab late in the month, until we found out why. So in the last few days of the month whenever we got a cab, we hung on to it for the whole day, refusing to pay until we got back safely to the hotel in the evening after work!

The Gezhouba system consisted of the usual dual VAX system with independent RTUs at each generator. In fact it was just a smaller GURI system, with almost the same set of software. One interesting wrinkle was created by the Chinese symbolic language. As by this time we had progressed to using our own design display generator, we could hard-code Chinese characters in pixel format into the display generator memory. This allowed us to invent what was likely the first all Chinese language HMI in a real time system. We first negotiated a deal to give them 400 symbols, but eventually we ended up with closer to 4000.

This was an advantage for quite a few years as no one else had anything that good, until the actual computer companies began putting in Chinese language systems. It even led to my being asked by a competitor company for a teaming arrangement on a bid to control the Ming Tan pumped hydro system in Taiwan. This was quite interesting because this partner company had their own controls division, which they ignored in favour of making us their control sub-contractor. On my visits to them at Vasteros in Sweden, I could not get a straight answer why they favoured us over their own kith and kin. It is an interesting example of how large companies can have strong rivalries within their own company!

The Chinese sent several engineers to us in Montreal for training and to actually do some of the software as well. We got a good reputation for being open and clear in training and that is a huge advantage in China, as they were determined to become able to do everything themselves. In all our later Chinese bids we always paid a great deal of attention to the training component of the bid and it paid off very well. By this time we at CAE had started putting Intel processor chips directly into our DATAPATH chassis, so we had our own computer board DMC (DATAPATH Micro Computer) to manage the RTU. This required that we recode the original GURI RTU software to accommodate this and other improvements that we had made. By this time we had moved on from using the DUSC based front-end to a system based on our new microprocessor (the same DMC that was used in the RTU). This is always a problem between systems as we invariably used the latest technology for all our systems, so we practically never made an exact copy between systems. Eventually many years later we licenced the design of our RTUs to a Chinese company, so that we could offer local content without which it was getting impossible to sell in China.

Our executives were at first rather reluctant to engage in the Chinese automation market. Initially they thought it was just one of my flukes that had landed us a project. By CAE's normal standards this Gezhouba system was rather small in dollar terms. So I started an internal campaign to impress the senior management on the size of the Chinese power market and even in the potential for flight simulators. I remember one particular meeting where I introduced some Gezhouba power plant executives to our management. It started off slowly but in the end our management were dumbfounded, but impressed when they learnt about the scale of projects in China. The Gezhouba dam was built by several thousand workers in record time. At another meeting the managers of Harbin turbine works, who were trying to team up with us, offered us help from teams of some thousand engineers. In those days China had started re-organising their enormous state owned companies in order to become self sufficient. So these companies were desperate to team up with foreign companies such as CAE to get independent work.   So we made lots of visits to various Chinese factories and ate lots of lavish banquets. I even learned to eat live fish at a very expensive restaurant in Wuhan. This was an expensive delicacy. Of course as time went by, the Chinese stopped bothering to team up with foreigners as their confidence grew and the whole market has become too difficult for foreign suppliers to compete with the local suppliers, except in certain very specialised fields.

This first Gezhouba system was followed by several full EMS system contracts for power grid control in different provinces in Hebei, Ningxia, Henan, Gansu,Sichuan, and the regional EMS system for the entire North West of China. In Sichuan we also won our first cascade control center to manage and optimise two hydro-electric power stations on the same river. At that time karioke had just been introduced into China and it was very popular. So in order to keep on the right side of the president of Sichuan Power during a negotiation dinner, I gritted my teeth and sang.

One advantage of working in China was that the Chinese did not use independent consultants the way utilities in America did. In fact during discussions with them they were quite puzzled with the concept. As they said to me "You know how to build systems and you have built several. So how can an outsider who does not build systems tell us what needs to be done". So in essence they used me as their consultant trusting me not to bluff. In China technical industries such as power are run by experienced engineers so they have a better understanding of what can be done. In North America too many utility executives are from finance so they have come up with typical audit type structures where consultants and other outsiders are used to minimise cheating.

Generally speaking we did not win systems in the coastal areas of China,while being more successful in the interior provinces. I did actively look all over China visiting Guangzhou, Tianjin, Guangxi, Guizhou, Zheziang, Shanghai, Yunnan,Sichuan, Henan,Hunan ,Fujien,Shanxi, Inner Mongolia and so on. On every trip, one could see China growing richer and more modern every few months. On my earliest trips there were virtually no cars on the streets, but millions of bicycles. In just a few years all the major cities were full of cars and traffic jams.

The Canadian diplomatic service in China were absolutely splendid in all the help they gave us over the years. On one long trip to Kunming in Yunnan I joined a team of other Canadian business men for a conference over there. This conference was superbly organised by just a single young Canadian lady foreign service officer who herded us all expertly around the area overcoming all sorts of local headaches with her infectious smiles! Obviously her Mandarin speaking skills were very good because the local Chinese jumped to her commands. The Canadian embassy itself was in a small building in the San li tun area of Beijing, but they could handle everything efficiently. As China opened up more and more I had to rely on them even more, especially as the local Chinese engineers became much busier and had less time to "stick handle" us foreigners. Even though we Canadians are from a small country, our ambassadors did not hesitate to argue our cases with the highest echelons of the Chinese government, making sure that we were dealt with fairly. According to Chinese history and propaganda the Chinese had developed a great fondness for Canada based on the exploits of our Dr. Norman Bethune on Mao's great march. A mention of his name always got us some favour and a few smiles.

On one trip to Hangzhou, I even rented a bicycle to wander the hill country and learned how to keep exact pace with hundreds of neighbouring cyclists. It is not as easy as it looks, for if you are out of step there will be a huge crash involving dozens of cyclists. One day I cycled to the top of a hill admiring the tea plantations, when an aggressive old lady collared me and insisted that I try her tea. So I went to her hut and had a cup. Then I bought some special tea for a small price. My agent told me later on that it was an extremely special tea which would sell for a fortune in Hong Kong. I took it home and tried it on friends and family, and I think it seemed to make them drunk. I even got collared for a walk-on part in a Chinese film. One day my Chinese colleague and I were having coffee in the hotel restaurant when a film director walked up to us and asked my colleague if he could "borrow your foreigner". So I got a minute or so in the film, though I never found out what it was all about. On a trip for the Wujiangxi dam system, my host in Changsha very kindly took me on a trip to see some very old archaelogical sites with Buddhist carvings. As he recognised my Indian heritage he thought that I could translate the ancient Sanskrit inscriptions, but of course I could not. It was also the very first time that I saw ancient Chinese mummified bodies in the local museum. Before that visit, like many of us from the west, I had assumed that it was only the ancient Egyptians who knew how to mummify bodies. It is amazing how restricted my knowledge of the world and its various civilizations was. I shall always be grateful for all that foreign experience that I got from my job while wandering the world. It has encouraged me to make reading about ancient civilisations and history one of my hobbies in my old age.

Getting a bid and its documents physically into China was a big headache. As usual there were several large documents, maybe of a few hundred pages each. Usually the customer required many copies of a complete bid, as they were sent out to various groups in different areas to evaluate the bids and give an assessment. In those days there were no laptops or easy ways to electronically send data. So we had to get the bid ready packed into boxes,in the factory, early enough to send the boxes by special courier. Unfortunately we were rarely ready in time so we often had to send someone at the last minute by regular air flights taking the boxes as checked luggage at great expense. I remember one effort that I made taking the bid documents for the Shuikou hydroelectric automation system in Fujian. I had to change flights for Fuzhou after an overnight stay in HongKong. Fortunately in those days the airport (kai tak) was in down town Kowloon and there was a connecting corridor with the hotel. So I was able to load all my many boxes onto a huge luggage trolley and man handle everything to the airport hotel for the night.

One error I made was not making a serious attempt to learn Mandarin Chinese, as I usually had a Chinese speaking colleague or an interpreter with me. To give you an idea how difficult it was to manage without the Chinese language, I can mention a long trip that I made alone. I got my colleague to fill a page of paper with the Chinese characters for useful words such as "airport","hotel","taxi" and so on. Then I took a long taxi ride from Yichang town to Yichang airport. On arriving at the small military type airport, the taxi dropped me off in front of a small locked building with not a soul in sight and then disappeared . So I sat for 3 hours on my suitcase and hoped for a plane, which fortunately did turn up and took me to Wuhan. At Wuhan I had no end of trouble getting a taxi and explaining the name of my hotel. Next day I returned to the airport and came for the plane to Guiyang. There were 5 identical planes on the tarmac. I didn't have a clue which one was mine, so I joined the rush into each cabin shouting "Guiyang" and getting a strong "Meiyo" (no) in return. I took a chance on the third plane and refused to move. As luck would have it, I was on the right plane and landed in Guiyang. The whole journey was quite worrying. Even simple things like buying a railway ticket became complicated in the early days. I once lined up for half an hour at a railway station for a ticket only to find out I was in the wrong line as I was a foreigner and had to go to a special foreigner line. I loved travelling by Chinese trains, as I could afford soft class. The train attendants always gave one a thermos of hot water and tea mugs, coming often to fill it up. At night they gave us the night bedding. While we waited for a train at the station there was always a special waiting room for first class (also called soft class) passengers, and usually someone came to welcome us on board. There even was music when the train left. For food on long distance trains I went to a common dining car and had great hot food cooked on the train virtually in front of me, actually on a stove on top of my table!

One project that helped us develop a complete backup hardwired control panel package in addition to the computer control system was for the Geheyan Hydro power plant on a tributary of the Yangtze river. This involved developing a completely independent hardwired manually operated panel based system of independent buttons,transducers and indicators, that could be used if the computer system failed. Eventually we found that such a hardwired backup design became quite important on many control systems, even in marine ship control.

Over the years I visited many sites where hydroelectric facilities are based in many parts of the world, but mostly in China. They are usually in beautiful but isolated hilly spots, so the travel there is long and interesting. Sometimes the newer and larger sites have the power house built underground. This is especially so in the cold arctic climate of places like James Bay in Canada. The general layout of a facility consists of the actual dam to hold back the water, the powerhouse and switchyards for HV power transmission to load centers. There is also a spillway to bypass the powerhouse and some times irrigation canal gates. On rivers like the Yangtze where there is heavy ship traffic, there are also a series of ship locks . On the Columbia river dams there are special fish facilities to allow spawning salmon to work their way back. So generally the control and management of all this can be quite complicated. If there are multiple dam sites on a river, the various facilities have to be coordinated to optimise the use of water.

The power house is usually set up as a large high roof building with all the generators laid out in a line. There is usually a ceiling based mobile crane system to be used when lifting any parts of the turbines and generators. As this floor is usually the cleanest floor it also contains the RTU and electronic control cabinets next to each generator. From the generators and HV power exciters the high power lines lead to the exit transformers in the open air. From here the lines go to step up transformers in the switch yard. On multiple floors below are the floors for the hydro-turbines and gates, a scroll case and the penstock that brings the water into the turbine. At the entrance to the penstocks of the dam there are trash racks etc to keep debris out of the power house. The water moves through surge tanks underground to smooth out flows. Finally there are gates to allow the used water to exit. In parallel with this equipment there are spillways with gates to spill water when there is too much. This can be spectacular. When the spillway is in operation at GURI, it throws the water spray so high it can wet the switchyard on the hill about 1 mile away. All these facilities have to be monitored, controlled and managed, including all the civil works, mechanical and electrical elements. All these elements have to be wired to the RTUs through termination cabinets and transducers, which makes the wiring a very big job as it is laid out on several floors and the station can be up-to half a mile long and 500 feet deep.

The computer control room is often on the top floor, with a viewing window giving a complete view of the generator floor. The RTUs are linked to the control computers by dual LANs laid out across the entire plant. In our design, each RTU could also be managed independently at a local VDU type terminal connected to the RTU. There is heavy mechanical vibration from the rotating machinery and potentially large high voltage electrical surge currents from various transformers. The lower plant floors are also dirty with dust,oil, etc. The humidity and heat can also be unbearable. So all our electronics and wiring had to be of the highest quality to withstand this environment. Our Chinese customers liked our RTUs so much they insisted that we always offer them on our bids even when we offered alternative cheaper Chinese RTUs to be more competitive.

The Chinese are great civil engineers, with lots of confidence. One system that I started investigating was financed by the world bank. It involved me doing some engineering consulting work to help design the control strategy before bidding the actual system. It was at Wanjiazhai in the north, just on the border of inner Mongolia. The project was a major water transfer scheme from the Yellow river to a water scarce region. The plan was to divert a portion of the river north at Wanjiazhai dam through a series of tunnels by pumping water over the hills and then allowing the water to flow by gravity into canals leading to cities and industrial sites. After use, the water was expected to flow back into the Fen river, a tributary of the Yellow river. If such a system had not been built, the entire city of Taiyuan would have had to have been evacuated for lack of water.

The pump stations were essentially like hydro-power stations working in reverse. To help the design along we came up with a full simulation to experiment with the control strategies to test what would happen. The design had to make sure that the pump stations worked properly in tandem so that there were no dangerous situations arising such as water flow back from the upper pump stations. The timing between stations was important to keep the cascaded pumping steady without surging. This simulation impressed the customer and the associated design institute in Tianjin. In fact there were several design meetings attended by consultant civil and mechanical engineers to vet our design. It led to my most bizarre flight into China. At the time I was busy in Montreal and could only justify a short trip to Beijing. So I flew out, taking all day to Beijing via Vancouver. Slept that night in Beijing. Gave a talk on the design next morning, and after lunch caught a return flight to Canada that very afternoon,helped by the fact that crossing the date line gave me some extra hours.

The area around Wanjiazhai was very dry and the people were poor. In fact many families lived in caves in the hills. There was a huge effort made by the government to plant trees by the millions on the bare hillsides. At each tree, they made a small depression to collect and keep any water as there was virtually no annual rain. This part of China has some very old portions of the great wall. For a long time in China, during the winter the government refused to allow any use of heat in office buildings, so even in major cities, we would have meetings in very cold rooms, wearing gloves and overcoats all the time. Fortunately there were no restrictions on copious amounts of fresh hot tea being served constantly. The "pipes" for the Wanjiazhai water diversion project were enormous concrete tunnels inside the hills, large enough to drive trucks through them. The pump stations were staged at different heights, enough to be able to get the water up the hill. The downstream plant included gates and weirs to divert water as necessary. There was a huge region where the population had been moved out to allow the building of a reservoir large enough to hold 2 months of water during which the pumping would be stopped. As this water would be expensive to provide it was only to be used for industry not agriculture. The Yellow river is so full of silt, the plan was to empty the system for only two months to allow flushing and cleaning the system. A radio system was planned for communication to all the elements with RTUs at each site. All the usual plant control strategies would be used within the pump stations. The communication and control cables within the tunnels and their associated electronics and power supplies had to be fitted in the tunnel roof and could not be accessed for the 10 months when the tunnels were full of water. The design had to be careful about overfilling or water surging.

The Yellow river is heavily overused and sometimes does not even reach the sea. I later visited the Yellow river conservancy commission in ZhengZhou for discussions on the Xialongdi dam, where I learned about these problems. In recent years China is building a cross China Canal to channel water from the well endowed Yangtze into the Yellow river. Until this is completed the Wanjiazhai diversion is still the world's largest water diversion project. Unless one sees the scale of all this engineering in such difficult countryside, one cannot truly appreciate Chinese engineering ability.

Close to my retirement time the Chinese authorities finally had decided to build the huge three gorges dam with three power houses. We did try to win the associated automation project, but it was such a high profile project our European generator/turbine competitors offered much better prices, even though we seemed to win the technical competition, especially as the design was based on our earlier work.

In the "greater China" region of Hong Kong, Singapore and Taiwan, we made several very large EMS bids but were only successful in Taiwan for all their regional EMS/SCADA systems.

One unusual opportunity happened because our sales manager had a good friend from Hong Kong who was a project manager in a major construction company over there. He happened to be visiting Montreal, so we showed him our systems and he was impressed. At the time his company was building a huge automobile traffic tunnel under the harbour in Hong Kong. He thought that his client would like to use a modern SCADA type system for the tunnel automation, rather than the conventional supply. I thought it would be a great opportunity to get started in a new traffic automation business, so I agreed to try a bid. There were two systems, one to manage the tunnel machinery for the support power and air conditioning, and another to manage the actual traffic flow and lights. To understand traffic automation better, I went to see the Traffic automation center of the city of Toronto to understand the technology. A highway automation control room, uses a whole wall of remote camera fed CRT screens to see what is happening. In addition electronic warning signs are computer driven to send variable messages to warn traffic to divert. Other than that it is mostly a simple SCADA type system, with traffic lights set according to the traffic flow. A traffic incident is registered by taking measurements of traffic flow based on sensors embedded in the tarmac.. Much of the required new incident predictive software came from California, where the new algorithms had been tested.

Our site visit to Kowloon and HongKong was very interesting. I think at the time most of the large construction cranes of the world were in HongKong as they were building the new airport and its associated highway system. Our tunnel was part of this highway. By now I was a frequent visitor to Hong Kong and even made a trip to check out the casino at Macao. One trip was even made just after a tropical cyclone driving through Guangzhou and Shenzhen to meet the China Light and Power (CLP)utility! We had to take a taxi all the way, as all flights had been grounded because of the storm. We could not delay as we had a meeting the next morning at CLP to discuss their upcoming EMS. For the traffic automation project, we were selected by the construction company, but once again our new executive management refused to cut our prices in order to get us started in a brand new business. My old bosses would have jumped at the chance to establish a new business area!

10. Islands and continents

As we had done reasonably well with control systems projects in Canada and Venezuela, our management decided that as a company strategy we should aim mostly for larger and more complex turnkey systems, especially as we were only marginally competitive on small standard systems. So our marketing and sales department were tasked with looking around the world for a suitable project. This was quite courageous on the part of our management, as large turnkey systems carry a great deal more business risk than independent systems. The prime contractor has to be responsible for a whole range of equipment, including much that is manufactured by other suppliers. A turnkey system is the epitome of systems engineering. Quite literally a turnkey contract means that one has to supply everything, including all the infrastructure to support a complete working system. So for a control system, it must include the power supplies, all the communications infrastructure, all the transducers and site modifications to the equipment that has to be managed and of course the control system itself plus all the required spares and training.

This is how I becam familiar with many parts of the continents of Eurasia, Africa and Australia/Pacific. Large turnkey automation system projects are rare in North America and Europe for the reason that the local power utilities have developed their networks over many years by themselves and are very confident and knowledgeable. In much of what is called the developing world however they need complete new systems and technologies right from the start. So they are likely to buy complete turnkey systems as they usually do not have the confidence to do project management themselves. This also serves the International finance authorities who prefer to finance complete turnkey packages as it keeps things simpler to have to deal with only one prime contractor.

The first trial project was a system formed by a Canadian consortium for a turnkey metro control system in Venezuela. The partner companies included the Montreal metro train control agency who were the metro experts, a prime contractor who would be the project manager and responsible for Venezuelan sub-contractors in charge of civil construction, a communications supplier for the radio and communications network, and CAE for the computer based systems. So I had to get to know the design of the Montreal metro control system, and make our part of the bid. However financial problems etc killed off the proposal. Though it did have the advantage of getting to know the Montreal metro system and its staff. Many years later we were invited to bid for their own system upgrade, but we stupidly didn't listen to their subtle communications about the computer that they preferred, and so we lost that project too.

The technology of metro control is fairly straight forward. Each line had a separate dual computer control system, driving a mimic lit up according to trains on track circuits. So the operators can visually observe the tracks and make sure trains were kept apart at suitable distances. All traffic moved in one direction only and so there was little need to switch tracks. There was registration for alarms indicating which station was involved together with remote operation of station loud speakers to give messages to passengers. Control panels,radio devices and CRTs within control desks were used for operator messages, reports and controls. There was a separate dual computer pair to monitor the traction power network, just like a normal power SCADA system. An interesting wrinkle on the Montreal metro was a system to monitor the air pressure on the train tires, as the Montreal metro used rubber tires on the trains (keeps the sound down)

After this attempt, we had another first rate opportunity for a world bank financed turnkey system for the West Pakistan power network. The utility was WAPDA (Water And Power Development Authority). The initial problem was to create a Canadian team with the best credentials but with us (CAE) as the leader this time. So we got together with other Canadian suppliers to form a team. The team members included a supplier for the Power Line Carrier subsystem, a supplier for the High speed microwave communications system, Ontario Hydro to be the overall systems engineer and power analyst, an architect engineer company for the civil works,site adaptation and construction and CAE for the computer systems and electronics. It was a most interesting training for me. In addition to the technical stuff that I had to learn, I also followed the commercial activities with interest. As a company we gradually became comfortable and knowledgeable with all the aspects of large scale international bidding. This included political background work, sometimes getting finance and understanding the local business environment. This was a significantly bigger and more complicated marketing effort than just supplying well specified computer systems. At first we had to persuade the client that we were a valid supplier with enough experience to be allowed to bid. After passing that hurdle, we had to go to Pakistan and find a local company to do the site adaptation and civil work locally. This was a crucial choice as a major part of the project was the site work and this was where our price would be the most vulnerable. I had to do site surveys to make sure our price had covered all the elements of the system. Later on after the bid evaluations we learned that we had provided the best technical bid, but one of our competitors had a much better price. It turned out that this competitor had used the same local supplier as us for the local works, and this local supplier had given our competition a better price than us. So in a sense we were cheated out of the project. I vowed that the next time I would control the entire price by doing most of the work by CAE directly. Eventually some years later this came to pass when we did the turnkey systems for the Egyptian power networks.

As mentioned previously, large bid such as this one, involve a great deal of engineering design work upfront, so making such a bid is a very expensive activity and can take several months and involve quite a few engineers. This WAPDA system consisted of two control centers, each one to control one region of Pakistan, and with the Northern center required to also supervise the two regional centers. So we had a dual VAX computer system with control consoles at each center. Every HV substation across the Pakistan grid was to be wired up to a CAE RTU. To do this an activity called site adaptation was to be done ie. the RTU electronics had to be interfaced with each potential power signal through an appropriate transducer. To find the correct interfaces we had to survey each site and check which transducers were the correct ones to buy and fit. The communications between centers and RTUs was through an older technology called Power Line Carrier, that used the actual high power cables to carry the communications signals as well. For high speed connections a microwave system that used a radio technology was to be set up, together with an attached telephone network. Backup diesel generators allied with battery-inverter power systems were required to power the control centers in case of the mains power failure. In this bid the actual control room buildings would be supplied by the customer, but in some turnkey bids even that has to be supplied. Finally before the complete system can be accepted, a long system test of several months has to be run using the actual facilities and staff who have to be given a long training as part of the contract. The clients usually had an independent consultant to evaluate the bids which had to describe the design in great technical detail.

In North America and Europe, utilities acted as their own prime contractors, and systems companies such as CAE only had to do the computer systems work. Historically turnkey projects were led by the civil construction company as the construction costs are usually the highest profile part of a project. These days customers are beginning to realise that it might make more sense for the control systems supplier to lead the team, as that is the most complex part of the job and usually also involves coordinating or making all the key technical design decisions. After all the civil parts of the project are usually quite standard and likely to be well performed. The systems part is where the new risk is to be found. Turnkey projects are very interesting, but expensive to bid. As some of these new clients were often relatively poor with shortages of foreign exchange, their projects were financed by international finance institutions such as the World Bank or the Asian development bank. When dealing in these markets it was crucial to have someone influential, who was a local, to act as our agent. This is a key appointment and will be the difference between success and failure. It is just not possible to sell systems abroad directly from a home base without an agent, as each country's business system and culture can only be negotiated by a local agent. So the correct sales strategy is to get the agent to guide the sales effort and to provide him with the technical support from home base. Any strategy that is based on restricting the agent too closely will probably fail.

Our agents in Pakistan were a very influential Karachi based business house. They had lots of influence with WAPDA executives, even including having a retired Pakistan army general on staff. So we were wined and dined in style on various visits there. Another bid for the control system on the large Tarbela dam on the Indus river also failed, when we lost again to the same supplier who had won the original WAPDA system. On account of my South Asian features the Pakistani customs officers at the international airport assumed that I was just another Pakistani expatriate from Canada. So I ended up having a long chat with him (in Urdu/Hindi) about life in Canada and he didn't bother our team with any of the usual customs harassment.

When it came to neighbouring India we also did not have any success in the power industry either. However after the many visits we made across the country we did create a good business with the Indian navy for machinery control systems, so the marketing efforts made our name known and paid off in the end. One of the activities I undertook in India was to find a local Indian company to be our partner, as Indian utilities insisted on local content. We did find a potential company who was excellent, but it all fell through in the end. They were very keen to partner with us at that time especially in nuclear power simulators, as by then we had invented a very innovative graphic process control language. India in those days was a strange mixture of advanced technology in computer software and very backward basic infrastructure. To impress us, they had organised an excellent computer demonstration of their software, only to have it all crash when the local power grid packed up. However long after I had retired I noticed that CAE had eventually completely purchased that very company and converted it into the core of CAE-India. International marketing and business development is a long slow process and requires a great deal of patience and only the most determined players stick it out.

My first marketing trip across India was as part of a Canadian government delegation that travelled across India trying to highlight Canadian power industry expertise to Indian power utilities. I got to know several executives belonging to other Canadian suppliers who were on the trip. Most of them supplied very large equipment such as generators, so I assumed that my small scope describing computer systems would be less important. Actually it was exactly the reverse. Indian utilities had not done any work with computer control, but their engineers were extremely excited by the work we had done and I had long discussions with them, almost to the extent that they ignored my fellow Canadian team members. The chairman of the Andhra Pradesh utility took a shine to me and we corresponded for a few years on various technical issues. A major organisational problem with Indian utilities was that generally their engineers were completely under the thumb of their bosses who were invariably administrators or politicians and did not appreciate technical stuff. Our Indian hosts took me to see the Bangalore control room,several substations and a 100 year old hydro power station on a deep river cataract. That hydro-station was so old that one would have had difficulty understanding how the generators worked. One of our team was a retired generator engineer who was thrilled to see them still working and he explained the parts to me in great detail. After visiting the main cities of Bombay, Bangalore, Mysore, Calcutta and Delhi, we even had a session in Khatmandu in Nepal. Nepal has a huge potential for generating hydro power but the only potential customers are in India. I did however get a view of Mount Everest and the Himalayas.

Many year later, well after we had successfully developed many full EMS systems in China, the Indian Power Grid corporation decided to implement large turnkey regional control centers across the country using World bank funds. Initially we could not meet their pre-qualification requirements for the first two systems in Delhi and Bangalore. However they changed their minds when it came to the next system for Bombay, which we bid together with an Indian partner company. Once again we lost to one of the original system suppliers. At this time I was also trying to interest both the Tata power company and the Bombay city power companies in selecting DMS(Distribution Management System) equipment from us. This involved doing some detailed design consultancy work to see how they could benefit from a DMS. As a result I had to survey several electrical sub-stations across Bombay city,including one enormous station right in the middle of Asia's largest slum. This Indian sub station was the best instrumented station that I have ever seen. Virtually every electrical and mechanical signal was measured and monitored.

Another Indian system that we came very close to winning was for the GMS (generation management system) for a cascaded hydro power system on the Kali river in the western ghat hills of India. The customer eventually cancelled that bid. I expect they used the money elsewhere. As you can see international bidding is very expensive but also has a low probability of success. Though when we did win projects it was very worthwhile. All these system bids were full of interesting custom design work during the bid process. For instance, the radio system required for the cascade was difficult to design as the hilly geography blocked the signals and required innovative thinking in setting up a relaying mesh network to bypass lost signal routes. Initially our CAE expertise was really only in computers, but the more we got involved in turnkey bids, I realised we had to learn about communications electronics and technology too. In North America this was always a separate supply, but to win a turnkey job we had to optimise the complete system. So I went back to my text books and started re-learning basic Electro-magnetic theory and the mathematics of communications theory.

The next big turnkey power EMS system that I got involved in was for the Greek power company that was centered in Athens, and needed a system for the whole of Greece. Again I had to make many trips to and around Greece finding sub-contractors and firming up the design. To help us in the bid we had a first rate local agent who happened to be a director of the local airline. He was so influential that one day he even held up my return flight as we were late and drove me right on the tarmac up to the actual aircraft door. Needless to say he made sure that I was upgraded to first class! He was also a first rate host and entertained us royally at the best restaurants. This is fairly usual when dealing with many international agents. They always entertain visitors very well indeed, and as they were usually fairly rich independent businessmen they did not grudge paying for fancy meals in top class settings. This of course made a welcome change from our usual company funded restricted daily travel allowances. By that time most international hotels and restaurants in the East were getting to be much more modern and lavish than hotels in the West, especially in places like Hong kong, Bombay, Jakarta, Singapore, Bangkok and Kuala Lumpur.

At the time some of our European competitors had finally started to use VAX computers as well, so we had to find some other advantage. In the meantime to lower costs on the Greek bid,we had come up with a computer design whereby we created a different type of VAX by putting together several small VAX computers tightly linked through shared memory. This gave us a computer that was as powerful as the larger VAX computers that would have been required, but at a much lower cost. At CAE we were fond of shared memory style inter-computer links. This was a consequence of the requirement in our simulator design where the shared X-ref database had to be accessed very fast by all the attached computers. Usually the industry standard in normal computers was to use inter-computer data links. This was too slow for flight simulators with their requirement for 30 to 100 msec calculation steps. It was just such requirements that had forced us to encourage the DEC company to invent the shared memory system for the GURI project.

When we eventually lost this Greek bid to a European competitor, one of the weaknesses of our bid was the perceived risk of this new multi-computer design. We failed to do a good marketing job in showing how reasonable the design was, and that we had used this design on flight simulators. Even the DEC company refused to give us much support as they were going to make more money on the standard VAX systems offered by the competitor. Probably we were also foolish in not understanding the strong political support that our European competitors had, as Greece was then negotiating to join the European common market. Computer companies like to sell larger, faster and hence more expensive processors while system suppliers prefer to keep costs down by inventing closely connected and cheaper multi computer systems.

The pricing required for this bid was typical of large turnkey bids. In such bids the client requires very detailed prices for each item of supply together with various escalation charges. This allows them to evaluate how much it will cost to buy spares later on, as these systems can easily last for 20 years or more. The prices for this Greek bid went into several dozen pages of data. Unfortunately we were running late in getting the pricing department calculations, so we sent off the sales manager to Athens with the technical documents and told him to expect all the prices by fax at the agent's offices as soon as we were ready. In the end we were ready only the day before the prices were due even bearing in mind that Athens in Europe was ahead of us in time. To avoid having a competitor getting a look at our actual prices, we sent them in a code which only our sales manager in Athens could understand and so he could hand write the actual numbers into the bid document. That night around 2 AM in the morning I got a frantic telephone call from our man in Athens that the numbers did not make sense, so I got out of bed and drove to the office and found some problem with our homemade codes, So in a panic I sent him all the numbers directly without coding as there was no time left in Athens to decode and finish the bid. In the end I believe our poor sales manager was hand writing in the last numbers while being driven in the taxi to the actual bid delivery point! I mention this episode to give one a flavour of the last minute panic in getting a large bid finished. This episode was not particularly unusual as bids always seemed to run as late as physically possible. Some of this was the usual delays as people kept trying to keep doing last minute updates to improve the bid. We also had to keep the final prices confidential, so to avoid any leaks we usually did not send final prices except at the last minute.

We became almost fixated with turnkey systems and even chased big bids around Africa in Benin and Kenya. One of the problems that the international finance agencies had in that part of the globe was the reluctance of western suppliers to work in such an unfamiliar part of the world. So they were delighted to encourage us to bid, but it did not work out and we lost both systems. The story was potentially more favourable with systems in Arabia, but they were well supported by the large European power equipment suppliers, so bids to the Saudi utilities did not work out either. They also had restrictive requirements about doing business in Israel which our management did not like.

We were still not phased by these losses and continued to visit potential clients in the East. In Thailand we bid for the national EMS at EGAT. Even though we had the low price, we found out too late that there was a favoured supplier who was selected even though they were higher in price. Our agent, who was a retired executive of EGAT, managed to get us visits to all sorts of senior people at EGAT but did not have a clue that the competition was fixed. In fact we got better information from our local Canadian embassy officers. Actually during my trips around South East Asia I discovered that Canada has first rate commercial councillors who were always very helpful to us. So I made it a point to seek their help in The Phillipines, Malayasia, Thailand, HongKong, Taiwan,Singapore and Indonesia. There were several bids in this area for systems during which I travelled all over the region. One of the highlights was my first trip to Sabah in Borneo . While there I managed to take some time off to climb Mount Kinabalu which at around 14000 feet is highest that I have ever climbed. An interesting aspect of the mountain climb was the dramatic change of climate as one went up. As this was near the equator, we started climbing in tropical rain forest, followed by temperate forest, then drier scrub plants finally getting to rock and tundra near the top. All the world's climates on just one mountain. After this exhaustive climb I had a great deal of trouble walking for a few days. I did get a chance to also take a boat trip to the pacific ocean where we were shown a sunken German warship from the first world war. The only problem in Borneo and south east Asia in general was the very high heat all year round.

In neighbouring Sarawak, some years later,I did a paid technical consultancy study for a potential DMS based in Kuching, but managing the whole Sarawak electricity network at the distribution level. . This involved doing a site survey of their gas-turbine plants in Miri and other electrical substations across Sarawak. I also studied in detail their utility power operations practices after carefully interviewing their senior staff. I discovered that it would not lead to an actual system purchase as they already had a SCADA system which I suggested could be economically enlarged with some modifications. The rest of Malayasia had started to also develop new systems as the country was growing very fast. We had bid for a number of regional systems and the national system, but at that stage we were not considered experienced enough. In the Cameron highlands we investigated a turnkey redevelopment of several small hydro plants, but the project kept being dragged out and I don't know what happened in the end.

In Indonesia I had to study and develop a proposal for three large turnkey DMS systems on the island of Sumatra. This project was brought to our notice by our flight simulator agents who were helping us develop a C130 simulator for the Indonesian air force. I made several site surveys and visits. The systems were to be installed in 3 provinces of Sumatra at Padang, Palembang and Medan. They already had transmission SCADA at each site, but were interested in Distribution automation. The site surveys showed that the area was quite heavily forested, so radio signals were a problem. The actual substations were quite modern and would have been easy to modify. During a break in the survey visit, my agent took me through the forest into the hills above Medan, which was a relief from the heat and humidity. Indonesia is a beautiful series of islands and well worth a visit. During this time in SE Asia there were many large forest fires that led to smoke covering even the big cities, including Kuala Lumpur. It was even difficult to see the Petronas towers, the then tallest buildings in the world. In the end the Indonesian client changed their mind about awarding that contract and decided to temporarily continue with their old systems. During this period I also visited the Indonesian navy in Surabaya to see if there was any prospect of business.

In a similar vein, we tried turnkey bids In Kazakhstan, where we could not get prequalified. In Thailand with another better agent we tried for a much larger modern system for MEA (Metropolitan Electric Authority) Bangkok, which we lost. We managed to reach the last two in a competition for a major bid for the Manilla electric company. However our competitor had a much better price, though we had a good set of technical meetings with the clients who liked our technology better. By this time modern control centers had very large LAN and communication networks and dedicated computers for each major function. In fact all the South East Asian systems of this era were very technologically advanced, especially as concerns the distributed computer network and fast computers. I would guess they specified much more powerful systems than the usual specifications of systems in the western world. As I have mentioned, they had the money and wanted the best and latest technology.

At about this time we were working on what were essentially turnkey type systems for the Canadian navy frigates, but we still did not have a power turnkey automation system. However our luck was about to turn. Out of the blue, a Montreal based Egyptian power device businessman came to visit us and suggested that we should investigate the Egyptian market. We were not very hopeful, but the offer of a trip around Egypt was too good to miss. So two of us went there and as our potential agent was very well connected, we had a full trip through all the tourist sites all the way down the

Nile river. We discovered that Egypt already had a modern national EMS system, and the main Aswan dam hydro facility had a good (Soviet)SCADA system.

Then as luck would have it, we visited the Alexandria Electric company (AEDC) on a Sunday when all the offices were closed. Our agent was important enough to persuade the chairman of that company to come and meet us. As at this time we had just finished the world's first full DMS system (as described elsewhere), I explained what we could do for them. Suddenly the chairman got interested and phoned around and forced his senior staff to come in and listen to us. This led to us offering to study their entire system and operations in detail and then designing an appropriate DMS for Alexandria. So that is what we did, studying their power network and substations including doing a detailed radio site survey as well. This study took a few months on site all around Alexandria. We discovered that Egypt had bought electrical power equipment from all over the world. There were substations from Russia, America, Europe, Yugoslavia etc. Nothing was standard, so we estimated that the site adaptation was going to be a big job. In addition to the main transmission substations we also had to automate each distribution point switching station and each separate distribution switch transformer/breaker box (called a Kiosk). As our normal RTU was too expensive for the smaller sites we would have to invent a new single board pole mount type RTU that was small enough to fit onto a power pole.

The SCADA radio system was a normal UHF system. Initially we thought that we would buy the radios, but the costs were quite high. At the time CAE had just finished developing a radio based FM system for transmitting visual signals on the international space station's arm. As CAE had very little experience in radio circuit development, we had hired a couple of specialists and they were running out of work. So to help them I came up with idea of integrating the radio IC chips directly into the single board pole mount RTU that we were planning. As by then radio design had become easier, because there were several appropriate IC chips one could select, I thought the risk was low and our price would be better. There was great deal of internal opposition, but by judiciously allying myself with the powers within our company I managed to push through the design. As there were only a few available frequencies , we had to optimise their use by scanning through a hierarchy of RTUs. In this way we managed to scan the entire system within the timing requirements for picking up changes.

For the mobile radio system we purchased the system from a local company. So for some time I had to once again read up a lot on radio communication technical details and became familiar enough to come up with a design for the network based on meshing and relaying techniques so that signals were scanned at a fast enough rate. One other consequence of this was that I became interested in designing a CAE spread spectrum (frequency hopping) radio system as well. That did go through a development stage but we gave up when we discovered that it was too expensive and did not go far enough. To test this system we tried it in the forests around Montreal, and found we barely could transmit even 0.5 kilometers in the forest.

One of our circuit designers even came up with a special modification to our DATAPATH I/O (Input/Output)cards to put in electrical transducers directly onto the card, to save the cost of separate transducers. The way we adapted the substations was to wire the high power switchgear through the top of their cabinets and then lead the wires to our RTU cabinet which was mounted in line with all the switch gear cabinets.   We had to cut or modify some of the local control wires that controlled the power circuit breakers, and install our remote control wires so that control could come directly from the RTU. This was a high risk as we were tampering with someone else's equipment. However it all went well and it took about a week per substation to adapt the control wiring for remote control. Each sub station building had to have the properly aligned radio antenna on the top of the building.   In some cases we had to get permission from the owners of a nearby taller building so that we could get a better radio signal. The adaptation in the kiosks was a bit more difficult as space was at a premium and the small pole type RTU had to survive the harsh Egyptian climate with no cooling. The actual physical work of site adaptation was eventually done by the power company's staff under our supervision. The guyed independent master radio tower masts were set up by us as were the actual control rooms, control desk furniture and UPS inverter power supplies. The software and computer hardware was a subset of the system we had invented for the St Louis DMS that I have described elsewhere. For the whole of Alexandria we designed 3 separate operations control centers, as it was a big area. We had to integrate data transfers and off-line administration between them too, so that the senior management could oversee everything from remote consoles at their head office. Even some of the maintenance centers in other parts of Alexandria had separate remote computer consoles for information. In all it was a very major computer systems and network integration effort.

One of the key advantages that this client hoped for was the facility to do feeder fault location remotely. In the existing set up, the control centers had to send staff out into the field looking at each kiosk to see which ones had picked up a fault and then manually isolating that section. By automating the kiosks our system would tell the control room which exact set of kiosks surrounded the fault. In fact we got carried away and tried to see if we could work out the exact fault location from analog current readings at the kiosk RTU. We even got a professor from the University of Saskatchewan to give us advice. In practice,the theory did not work out so we gave that idea up and stayed with isolating the fault to between kiosks. Eventually on the Taiwan system we came up with a better isolation scheme for remote feeder fault location and automatic isolation but that was much later.

Finally we had our long efforts and patience in South East Asia rewarded when we won the world's very largest turnkey power automation project. This was to automate the entire provincial distribution power grid of Thailand, outside Bangkok. Discussions started some years before the actual bid, when a large group of senior Thai engineers visited our factory, having heard about our unique DMS system for St. Louis. They were determined to only allow a very short list of particularly well qualified suppliers to bid. In the end only 3 companies were allowed to bid.This was by far the largest power automation contract in the world and even included the actual control room buildings. It was world bank financed and very hard fought. In fact after the first round of bidding, there was a great deal of local politicking for advantage. This bickering even made head lines in the main Thai newspapers. In order to get the correct winning price I analysed our losing bid for the previous MEA Thailand bid and did some judicious guess work to get the winning price to just below that of our main competitor. So in a way losing that smaller system did help us win the much larger system. There were multiple control centers, one in each province of Thailand. The contract included a training scheme for a very large number of Thai engineers. As it was a turnkey system it included the supply of the radio system, the RTUs, site adaptation at the substations, and all the secure inverter power supply systems required. This was a final vindication of all our international marketing efforts especially in the DMS systems that I have described elsewhere.

At about the same time our patience was also rewarded in Taiwan where we won system contracts for all the regional power control centers across Taiwan. This win eventually led to winning the first distribution feeder automation contract on the island as well. In this system we had to invent new computer coordinated automatic switching of feeders when faulted in addition to the usual DMS applications.

My efforts in Israel were not successful. As we were in Egypt at that time, we just flew in and visited the IEC (Israel Electric Company) to talk about their EMS system. It was just on the off chance that we might find something useful over there as it was barely a half hour flight from Cairo to Ben Gurion airport. In many parts of the world potential clients are usually delighted to meet engineers from North America as it makes them feel part of a world wide network of automation engineers. I had expected a great deal of security checking etc, on a flight from Cairo, but there was literally none. One just got on the plane and off we went! The Israelis were very friendly, though we lost to a more experienced company, but it got them interested in a huge turnkey DMS for the whole of Israel. They really did want us for that system, but our top executive management had changed at that time and refused to make any price concessions. While in Israel my colleague and I thought we would drive from Haifa to see Jerusalem. We took the west bank road through what was effectively settler territory, being quite oblivious to the armed troops etc. Afterwards when we returned the car to the hire firm they roundly criticized us for doing such a dangerous thing with their car. On leaving Israel, they have a system for interrogating all passengers to check if there is any danger in letting us fly. When I told them about visiting the Israel control center they immediately became suspicious and kept questioning me more and more. The Israelis hide critical infrastructure so that no one knows where it is. That control center was hidden in a normal suburban district and from the outside looked just like another tract house. In fact we drove around the area for a long time before we found the place, as it was well camouflaged.

By now we were quite popular in Egypt and one winter I had to survey all the substations in East Cairo for a DMS over there. It seemed that most of the electrical power equipment all over Egypt was similar to the substations in Alexandria. We also won a set of very large turnkey DMS systems for the canal Zone power company based in Ismailia. This was the largest electrical region of Egypt and included the Sinai peninsula. The basic design was the same as Alexandria, but the systems were much larger.

CAE did win several smaller SCADA/DMS/EMS systems in other parts of the world, but I was generally not much involved in those systems. These systems were installed in Honolulu (Hawaii), Espoon (Finland), Rekyjavik (Iceland),Maribor (Slovenia),AGL (Melbourne Australia). We did not at first have any success for other SCADA bids in Australia at Rockhampton in Queensland and Perth in western Australia.

After we had finished the big DMS for Union Electric in the USA, for a short time we were quite unique as the world's only experienced full DMS supplier. This is what had led to the systems in Egypt, Thailand and potentially in Israel. It also led to our first big system in Australia. The city of Brisbane power company (Energex) in Queensland, sent their chief designer to check us out. He got a very good reference from Union Electric and so awarded us an enormous full DMS for the region. This was, why I made several trips all the way to Australia. I also teamed up with them on possible joint Indonesian bids as the Australians were keen to expand business in Indonesia. During my spare time in Brisbane I used to rent a bicycle and cycle all along the river for miles. I was interested to learn that one of the key reasons how they justified such a large DMS and the associated distribution automation was that it allowed them to quickly isolate the electric system just before tropical storms swept through Brisbane. These storms were common in summer and did a lot of damage. The savings from just protecting the power network and quick automatic restoration was more than enough to pay for the system. This system had a full range of Outage management applications to quickly locate, isolate and re-install electric power all over Brisbane. Distribution management is different from most transmission SCADA as most of the electrical plant has not been modified for remote control. So the only way they have to manage outages is by physically going to the site and manually switching breakers. To do this they have to go through a secure administrative procedure for "outage management". It was this software that we supplied in addition to our usual full graphics maps and SCADA. They had automated some key switching sections, so we came up with software to prepare lists for automatic sequence switching when in a hurry.

This system led to the Queensland provincial HV transmission electricity utility (Power Link) also coming out with a full transmission system EMS specification which included every single possible HV power software application. After I analysed the bid I initially thought it was too complex for us, but the then company president insisted I bid it, so we did and we won. This seems a good place to describe such a full EMS system. It included all the usual hardware at the master station with dual main computers, master station LANs, control consoles, front end microprocessors, inter network gateway processors and field RTUs. By this time we had progressed to DEC alpha 64 bit computers as they represented the most powerful available DEC computers. In addition, by now our control desk work stations were full graphics work station units (usually Silicon Graphics). The programming language had also been changed to the C language, with the UNIX operating system. All these changes were made mostly to keep our systems up to date and generally with the most powerful equipment on the market. We had by now pioneered the use of OPEN standard communication software and POSIX compliant operating systems within Real Time systems. We used the most common standard Relational data base systems too. At first we used SYBASE, but eventually we were forced to change to ORACLE as that database system had become the most popular. We stayed with Ethernet LANs, as somehow Ethernet always had the fastest and least expensive equipment, even though other technologies such as FDDI, ATM and other Token ring buses had been invented.

Managing a high voltage electrical system is a very technical activity, but the electro-magnetic science is well understood and is susceptible to advanced mathematical analysis. In summary a power network consists of generating stations, which could be fossil powered, or nuclear powered or hydro powered. These generator stations were connected to load centers such as a city by transmitting HV AC power over long distance transmission lines running at very high voltages. Interconnecting these lines and switching the voltages are electrical substations with transformers,breakers, capacitors and voltage devices. All these stations are connected to the control center through RTUs and an overlaid communications network which could be wired or wireless. As power cannot be easily stored, the network has to continuously generate the required power instantaneously and so has to be monitored by computer to make sure the power devices keep to the required frequency while maintaining the required safe operation of such a complex network of devices.

So EMS applications software falls into 5 categories. As part of a normal SCADA system, monitored data is aggregated and shown in various historical forms and reports after calculations are done with the data. All major power devices are wired up to the RTUs so that control operators can safely do remote control actions at the control centers, after monitoring remote events and alarms. The power operators have to continuously monitor and control the generation of electricity and the transmission of the power. So there is a set of software based planning tools, a set of generation management software applications and a set of transmission management software applications and finally an operator training simulator using the same control software working against a computer simulation of the network.

The planning tool programs include:
Network Operations Scheduling, Interchange Transaction Scheduling, Outage Scheduling, Load Forecast, and Hydro and Thermal Unit Commitment with HTC co-ordination.
The network transmission management software includes:
Weather and Load Forecasting. Network Security functions including State Estimator (essentially a real time power network simulation), Pre-switching Validation, Contingency Analysis and Remedial Action, Reactive Reserve Monitor, and Security Constrained Dispatch. Network Optimization tools include Optimal Power Flow, Penalty Factor Calculations, and Volt-VAR Control.
The generation management software includes:
Automatic Generation Control (ie. managing network frequency),Economic dispatch,Voltage and VAR control and dispatch,Production cost monitor and reserve monitor.

To get an idea about what these programs do I shall describe a few in a short form. Power flows across a power grid network in strange ways and it is difficult to know what the actual current is in any particular segment of a network at a particular time. However it is possible to make a close guess using a program called state estimation that uses a form of Kirchoff's laws to make a first guess followed by reworking the calculations through closer and closer iterations, until the loose ends seem to be small. This is considered the best estimate of what is happening. Then the operator can run these sorts of programs after imposing potential faults or shortcomings on this network simulation to see what would happen. This leads to a set of possible dangerous situations that the operator can then try to avoid. If the control operator is not careful, the electrical system will have a blackout caused by a cascading set of protection devices that will automatically shut off the entire power system. This activity is called network security monitoring. More sophisticated programs such as optimal power flow and security constrained dispatch work out the best and safest allocation of power flow according to the existing situation on the grid.

When it comes to monitoring power generation, the AGC and economic dispatch type programs make sure that the system frequency is kept within close bounds by increasing or decreasing generation. If the frequency falls too low the electrical generators will automatically cut out and there will be a blackout. The economic side of the calculation chooses which particular generators are to be used and in which proportions in order to minimize costs. The unit commitment and other scheduling type planning programs do the same kind of optimisation but with a long term horizon. The weather plays a big role in planning safe operations and so does the forecast of how the electric load will vary during the day, so there are programs to take this into account. The Volt-VAR control strategies are used to minimize losses arising from the inherent characteristics of electro-magnetic systems to fluctuate according to the inductive or capacitive effects. Reserve power has to be available so that the system can immediately respond to major changes in demand'

One could continue to give more detailed descriptions of what each software program does, but the program names are quite descriptive and it would take too long, and may be boring. Suffice it to say that electrical power engineers and scientists have developed these algorithms and strategies over many years dedicated to managing large electrical networks. Most of this type of work used to be done in an off line manner using paper and pen. It is only in relatively recent years that these strategies and analysis have moved to computer software. At first these programs were run off line. Now the software can be run directly on line as modern computers have become powerful enough to produce results in real time. At CAE, since doing the Queensland, NW China and Hydro Quebec systems, we have probably implemented the widest range of such software. There are also other software packages for power system black start (A large power network cannot just be switched on, one has to gradually bring power back in stages), training simulation and so on. The power system application software for distribution networks can be similar in concept, but since there are very few remote instrumentation and control facilities, predictive software is not that common or useful.

Often all these software applications have to be very customised as each power system is quite unique. Not all power system operators use the entire range of such software, preferring to use some well tried heuristic approaches. I have even been in control centers fitted out with a full set of software applications,which were never used. It can be a fashion item for some utilities. I think the biggest problem with these applications is that for them to work correctly, the entire power database with all its parameters has to be kept continuously and completely up to date all the time, and that is a lot of work for a busy operator. In addition power networks have been built over many years and are full of strange local peculiarities, so generic software cannot be usefully used directly. An enormous amount of customisation is required for each network and each power generator.

After Queensland, the customer was so pleased with their system, they even tried to sell a copy to the Tasmanian utility,asking us to supply the hardware only. However that bid did not succeed. Our name did spread around the area and I even had the opportunity to bid a small DMS system for Auckland in New Zealand, but the cost of travel there and back was so great it was not worth it for a small system. Sometimes for small systems it makes sense to use a local distributor to licence and sell the system, but somehow we never got comfortable doing that.

11.The deep blue sea and sky.
The Sea

Sometimes good fortune works in strange ways. We at CAE created a brand new business from nothing, that earned well over 1 billion dollars in revenue in less than 15 years. This is quite remarkable in the systems business, which is brutally competitive. Here is how it all started.

One afternoon I was waiting to see my boss, who was temporarily out of his office. I found some technical looking documents lying on his side table, so I started browsing through them to kill time. They happened to be an RFP (Request for Proposal) from the Canadian Navy for a prototype test system to prove a new ship control concept. When my boss returned, I asked him what it was all about. He told me that he was about to throw the document away as the military flight simulator department was uninterested in following it up as they said it was not a flight simulator! I laughed and told him it looked like a small control system and should have been sent to me, not military simulators. He was delighted that I was ready to work on it as otherwise he would have had to disappoint the sales department.

The Canadian Navy is a rather small navy, but is extremely advanced in technical knowledge. Their officers spend a lot of time and effort designing and developing advanced methods for naval and weapon engineering systems. Unfortunately the government does not give them adequate funds to actually build systems, so many of the concepts remain as paper designs. One such design effort was to develop a distributed computer based concept network to monitor and control weapon mission systems on board ship which they called SHINPADS (Ship Integrated Processing And Display System). This was very unusual and innovative for those days. They gave development R&D contracts to some Canadian military equipment suppliers to develop the required military specification computers, full graphics displays and distributed Electronic Bus interfaces. When the engineer officers of the navy saw this design, they thought that they would also piggy back onto this weapon system design and use its elements to operate and control the ship machinery (propulsion etc) using the same distributed concept. So they called it SHINMACS (Ship Integrated Machinery Control System). The document that we nearly threw away was an RFP to develop a prototype SHINMACS to evaluate its proof of concept.

The idea of this RFP was to use the exact same electronics (computers,displays, bus) to prepare a single machinery control console, and then use it to drive a computer simulation of the ship and its machinery. The navy would then use this equipment to see how they could use it on an actual ship. So I put a competitive bid together which we won and the Navy then awarded us the contract.

The system consisted of a VAX computer on which we ran a software simulation of the propulsion machinery of an existing Canadian navy destroyer. This was interfaced to a CAE RTU that mimicked control electronics interfacing with ship machinery. The customer provided the SHINPAD naval computers to run the shipboard network and the control display generator. Using this equipment we produced the application (SCADA type) software and display software. Once the system was ready, the navy sent down a team of sailors and actually ran a whole week keeping the system running round the clock, as if on board ship. They were delighted with the results. It proved to them that it was feasible to control a warship exclusively using modern electronics and computers only. This approach of doing such an expensive trial was a surprise for me as I had been used to the Industrial world where customers just specified full systems without doing any prototype work. One element of the design that impressed me were the control display pictures. These had been carefully developed by a specialised human factors company. They allowed the ship engineers to access the various controls in a very intuitive way and they could then easily trace faults etc. I suspect these displays were the forerunner of control displays that are now used in the "glass cockpits" of modern aircraft. As such they are a much better and more intuitive approach compared to the simple line diagram and alarm list approach that had been used in the power industry. This experience then allowed us to design similar HMI displays on future power industry control systems too, especially for generator and machine control pictures.

At around this time the Canadian government was in the process of contracting with shipyards to build new Patrol frigates. By specification these were intended to have the very latest machinery, weapons and the SHINPAD computer systems. Unfortunately as the government departments are very conservative in technology matters, they insisted that the ship machinery controls had to be done by an experienced supplier using conservative equipment only. So there was no way for us to be allowed to bid for the shipboard controls, even though we were the most experienced Canadian general control systems supplier. We decided to team up with both competing prime control system teams. They were glad to have us on their team as the contract terms required a large amount of work to be done in Canada.

The prime contractor for the weapon systems was an experienced American defence contractor who had done similar work for the American navy. The shipyard was Canadian. As the new machinery control system was supposed to be a software system running on computers this sub-contract was allocated to the weapon prime as they were comfortable with software. In those days shipyards were strictly mechanical and heavy electrical engineers. They could not have properly supervised electronics and software work . It is worth remembering that this would be the first time that a warship was going to be controlled by computer software. Even the experienced ship control suppliers were offering development systems, that had never been to sea. The two competitors were an American supplier and a British supplier. As a potential sub-contractor to either, I had done a detailed site survey on board the Canadian navy's advanced destroyer HMCS Algonquin when it docked at Halifax on our East coast. The navy sailors were most helpful and showed me all the details on board ship, including a peculiar pneumatic/hydraulic control system. As an aside, years ago I had seen similar pneumatic systems used in England, but in the end this was to only be an intermediary technology in between mechanical and full electronic control. This happens fairly often in the technology world where orphan technologies are tried but get stranded and replaced by better (or maybe more popular) techniques.

As one can imagine there was a great deal of politicking to winning the various contracts. The weapon systems prime was used to doing American government business in which political influence is most important. They initially favoured the American machinery control prime, but were quietly told to favour the British supplier, as we (Canadians) were also on that team . As part of the British team I went to the UK to negotiate for a share of the contract and view the new control system that they were going to offer. It was virtually the same in concept as our own system. Through the back door, our Canadian navy engineer friends told us that they wanted our British prime to offer an option for a design based on SHINMACS as well as their own British system. However when we passed on this information to our potential British prime they refused to be persuaded and insisted on offering only their own design system. This made things impossible for us to offer the SHINMACS as an option. Our Canadian navy engineer friends were not to be blocked. They arranged for the British bid to be rejected as it was too expensive. In spite of all the good marketing intelligence that everyone gave the Brits, they falsely thought that they had a lock on the contract and they refused to reduce the price. So our team fell through, leaving the American control supplier as sole bidder. As by then we had made a very competitive design based on our work with SHINMACS, we were asked by the weapon prime to put in an alternative bid, as they understood very well what the navy really wanted. Our bid was very price competitive with the other American bid, and as the Canadian navy really wanted SHINMACS, the weapon prime selected us directly.

The British company's failure was symptomatic of their arrogance and failure to pay attention to marketing intelligence. I saw some of this when I was trying to get a reasonable share of the system work on their team. In those days British prejudices led them to believe that we, as a Canadian supplier, were only capable of doing simple work such as beating metal into control consoles. This was in spite of the fact that we at CAE made many more complex systems than they did. One of their marketing guys noticed this but was not influential enough to persuade his managing director. One consequence of losing this Canadian contract was that a few years later that British company had to close for the lack of new business. This in spite of their long and glorious history in control systems. It taught me a lesson that I never forgot about treating every significant item of business as a life and death matter and to never give up

A naval ship such as HMCS Halifax (the lead ship in the Canadian CPF contract), has a complete range of propulsion and facility management machinery in addition to the weapon systems. The ship is divided into two main groupings ie. The platform and the weapon suite. Our job was to supply the systems that manage and control the ship platform for which we invented the term IPMS (Integrated Platform Management System). The propulsion suite consisted of a cruise diesel engine,and a primary set of gas turbine engines for high speed operations. These drove the variable pitch propellers through a set of propeller shafts. The different engines drove the shafts through a complicated set of connecting gear boxes. The fuel system consisted of metered pipes,tanks and pumps.The electrical system of the ship consisted of the diesel generators with electrical distribution wiring throughout the ship through electrical switching distribution points and transformers. There were a set of ship "hotel" services machinery such as the clean water and black water systems, the air conditioning systems etc for the services to support the crew. At that time there were also two other ship wide systems for voice communications and for damage control. These were a separate supply by others, but in more modern designs are often integrated into the platform management LANs as well. Even the ship navigation systems can be integrated into the IPMS as another node. So essentially a ship is just like a complicated "mobile" power station, and so it was not difficult for us to get to grips with it.

The weapon prime had undertaken to build a complete land based system and simulator for the ship that was to be used to test the actual ship equipment before it was installed on board ship. So they required that we also do the same and develop a land based test facility for the IPMS. This facility included a complete system copy and was connected to a computer that ran a detailed simulation of all the ship machinery. To keep the equipment as accurate as possible, we even had to interface between SHINMACS and the simulation computer using physical SHINMACS RTU to test RTU electrical connections in case of timing issues. It was also to be used to test out new replacement or repaired parts before these parts were installed on board ship.

One of the peculiarities that we had discovered during the SHINMACS prototype development was that the SHINMACS military computer devices were too obsolete and expensive. We discovered that we could make a more powerful and cheaper system if we kept all the applications software within our own RTU and microprocessor equipment, and use the specially designed SHINMACS computers purely for interfacing to the bus. During early technical discussions the navy officers were horrified at this waste and insisted that we throw away all those specially designed expensive military computers and keep only our own modern DATAPATH M equipment, thus saving a huge amount of money while producing a more powerful system. The American prime contractor was very suspicious that we had always meant to discard the original design, but it was not my intention at all. It just was a better approach and the navy chose it. So we unfairly got blamed as very conniving company. Another peculiarity of this contract was the enormous amount of bureaucracy involved in a military contract. This is quite unlike an industrial type contract where things are done in a more practical way. I first saw this effect at the very first design review meeting with the prime contractor. They insisted that I explain every major design decision such as why we chose various parts of the system. How could I tell the "august" members of the design review that I had chosen the various system elements because that is what I already had and it was cheap enough! They insisted that I do long design calculations to "prove" it was the best design etc. In the end they gave up harassing us with bureaucratic procedures and left us alone to work any way we felt comfortable. This is the one element of working on government programs, especially US government programs, that I find most annoying, but I expect government employees are never satisfied unless they see a mass of paper for every bit of equipment .

The way the contract was written was that for all practical purposes we were a turnkey control supplier. We had to also supply all the required transducers and interface with with all the machinery, taking all the risks that we would supply a complete working machinery control system with nothing left for others to do. There would be a total of 12 shipsets, so it was an enormous contract, spread over many years. As these were military ships, in those days we had to militarise all our equipment including the electronic PCB cards. Militarizing essentially meant that the equipment had to stand up to heavy vibration, shock (from weapon firing), water proofing and heat. To understand this new environment and design the field testing approach, our draftsmen, manufacturing and test engineers had to learn a new way of working. Fortunately the naval test establishment was near our factory so we could do all our prototyping and hammer tests close by. Militarizing meant that the cabinets, chassis and consoles had to be made with special steel structures and paint, including shock absorbing springs. The most complex and delicate equipment that had to be protected were the CRT displays, that had to be specially made by MIL spec suppliers,and then installed in the shock proof consoles. All the PCBs were also militarized in our factory, with special cooling facilities (no fans) and carefully assembled to withstand shock and water. The QA was also run according to the MIL specs, so then CAE became a full MIL spec supplier for the first time. We renamed our marine electronics series DATAPATH M. The distributed network consisted of 3 independent LANs which were to be laid on the port, starboard and bottom of the keel. At suitable points the microprocessor based RTUs connected to each LAN. At that time the main LAN was initially the SHINPADS LAN , later we used the CAE proprietary RS422 type copper LAN system for the RTUs and the master microprocessors. Many years later it was all upgraded to fiber optic Ethernet type LANs. Even the panels, instruments and push buttons had to be special MIL spec devices.

Eventually updated designs moved to LCD type displays as they were more reliable than CRTs. One very special control console and RTU combination was called the LOP (Local Operating Panel) and was to be used very close to the actual turbo machinery in the "dirty" engine spaces. So this had to be manufactured to even more stringent environmental standards. We even had a console on the bridge as well as the operator consoles in the machinery control room. The bridge consoles were used for remote operations of the machinery, including the steering gear and to also eventually interface with navigation type equipment. Eventually we had to redesign smaller RTUs for ships, these RTUs being based on militarised versions of our pole mounted RTU. This was necessary to avoid having to bring a lot of field wires up to the larger RTUs. It was easier to spread the wiring by having more RTUs per ship.

At the time the new industrial style aero-derivative gas turbines (LM2500) planned for the ship did not have an engine computer controller, though the turbine manufacturing company had planned a new design digital controller. I discovered that by designing a few special PCBs we could actually build an electronic controller directly from within our RTU electronic range. This would have been significantly less expensive than buying from the turbine supplier. This suited the navy as our design was more powerful and minimised the number of special electronic circuits that were needed, hence saving costs on long term maintenance. That was how we came to design and deliver the very first computer based direct engine control module (ECM). The correct technical term is FADEC (Full Authority Digital Engine Controller). Some controllers have no other seperate hardware backup, but we didn't behave so foolishly and had safety switches too. Today it is fairly common to have computer based direct digital engine controllers, but at the time it was an industry first and replaced the usual complicated and expensive electromechanical-hydraulic fuel and engine management controllers. I explained how we would do this to our executives, and as usual they took the enormous risk and let me go ahead. It was quite a surprisingly short meeting, where I also described how we would militarize our electronics, especially the graphics system. I had expected that our senior executives would have been most reluctant to take on so much risk, but they took it all in stride, drank their coffee, told me to do a good job, then walked off!

The ECM had to include vibration monitoring PCBs, using the then new DSP (digital signal processing) IC chips. It also had to interface with the mechanical fuel control levers, so that the ECM computer would properly meter the fuel flow according to computer fuel schedule lists. An emergency shutdown, cutting off fuel flow had to be designed in. Control logic sequences had to be built in to do all the preliminary activities such as engine wash etc. As this was going to be a new untried approach, we had to design it in careful stages. At first we teamed up with a specialist gas turbine expert company who specified how the equipment should function and helped us develop a computer simulation against which we would do the initial tests. Once we had carefully done these tests we had to take our prototype equipment to a special US navy test bed in Philadelphia where they had a land based, specially instrumented actual LM2500 test engine and go through all the actual engine runs. Finally once we were satisfied we had to do all the test runs on board the actual first ship to be sure that the engine controllers worked properly. We also introduced modern computer based health monitoring software for the machinery, running in our microprocessors. Health monitoring for turbo machines is mostly about vibration, hours of use at various speeds, lube oil condition and so on.

At first we wrote our software in Intel PLM, but eventually our marine software was moved to the ADA military programming language or the C language. The SCADA part of the software was almost an exact conceptual copy of our usual Energy systems software, but based on distributed microprocessors, So a lot of the control software was actually loaded into the nearest intelligent RTUs, rather than in the master coordinating microprocessors. On such a distributed system, the software is kept as close to the actual devices as possible, with the top level microprocessors doing more coordination rather than the actual control applications. Each graphic console microprocessor included the detailed graphic display diagrams and HMI in addition to producing various reports. In addition to spreading the load such a design helps to make the system more reliable and resilient. One unique element that we introduced was an on-board training console and simulator that could be used by the operators during quiet times at sea. This simulator used the same electronics as the main system. Again this was a similar idea to the simulators we had done on power control systems such as GURI.

While making site visits to check our work at the shipyard in St John, I was most impressed to see how modern shipyards build the hull sections inside a closed factory, and then finally assemble the sections outside. Our field site manager had a great experience learning about ship board work. All the suppliers had a good relationship with each other. I remember them joking about how poorly the new toilets worked on the ship. After the first ship was ready, it was taken by the new crew for sea trials, and we all crossed our fingers and hoped for the best as everything depended on our computer systems. Initially there were problems, including a test where the ship computers "bombed" and manual overrides had to be used to stop the ship. This news even made the daily papers! However it was all fixed and the Canadian navy was very proud of their ships. To help us sell our equipment in future years they even sent the frigates on courtesy calls all around the world so that other navies could view their equipment. Years later during mid-life refits, they came back to CAE directly to upgrade all the control equipment.

In parallel with the Patrol frigate contract, the navy decided to update their tribal class destroyers as well. Once again they selected CAE for a similar system. There was another prime contractor to whom we were responsible. The gas turbine was from another manufacturer (GM-Allison), so we had to re-design our ECM with the crucial help from our Gas turbine expert partner company. By now we were getting quite confident with the naval environment and the work required on board ship. From a business point of view the projects were a huge success, as all the development work and cost overruns were charged to the first ship set, leaving the follow on ship sets to bring in a reasonable profit.

As these were the very first naval ships to use software based computer control of machinery, other world navies began to call and talk to us about their plans. One of the first was the Israeli navy, who we went to see in Washington where their ship design team was working with an American design team. Their chief machinery designer was a retired Israeli navy engineer officer who was very keen on our design, especially as it reduced the requirement for a larger crew. Generally it seems that crew size is not that important on military ships as people are needed during action to replace injured sailors or do manual work arounds. Commercial ships on the other hand want the smallest crew possible, so that is why full automation on merchant ships was attempted long before. The Israeli boss, an admiral, was also enthusiastic as Canada was well regarded in Israel. However the working level engineer officer was most suspicious about how well all those computers would work in a dirty engine space. So we had to spend a long time reassuring him. Many years later when I was talking to the engineers of the Israel Electric Company (IEC) about their EMS system, these contact references were crucial as an introduction, especially as the IEC engineer was also a retired navy engineer. So that is how we won our first international system for the Israeli navy SAAR 5 corvette. This was also around the time of a contract for US navy mine hunter ships.

After this we began looking further afield, especially in India where the Indian navy was building a large fleet. We became quite close with the engineer officers at headquarters in Delhi. Our agent was a retired Indian navy admiral, and he was most helpful in getting us to meet their highest serving admirals. He arranged dinner parties for us with the top 4 admirals at their private club. He even got us invited to several design meetings with naval staff at the Mazagaon dockyard in Bombay. As a side benefit of my Indian features, the security staff at various military establishments assumed that I was a local and let me enter their military facilities quite easily. My other Canadian colleagues had to go through a huge rigmarole while being checked in as foreigners. The Indian navy were extremely enthusiastic about our system, but Indian government rules were very stringent about buying equipment from abroad. An essential requirement was that we had to have a local partner, and we spent a long time traipsing all over India talking to potential partners. We got quite close to two very large Indian companies, but somehow could not quite close the deal. Eventually many years later, we gave up and just started our own Indian company with which we began winning contracts for frigate type ships. Over the years I spent many days talking to high level Indian company executives, but they were too conservative to sign up with us. One of them tried to interest us in another military system for a tank and artillery simulator. So I came up with a design and we presented it to the chief artillery general in Delhi. However as it would have been a development contract for us, they preferred to go with someone who had already built such a simulator for the British army. Strangely enough many years later CAE, in an expansionist mood, actually bought this very same company as it had fallen on hard times.

The key navy that we were interested in of course was the US navy. We first started by talking to their David Taylor development labs who gave us a big development contract for a new design based on a different microprocessor family using VME (Motorola based)cards rather than Intel cards. This type of system was eventually targeted for the US navy's LPD ships. So in this fashion, our marine business grew in leaps and bounds.We had early contracts from the UK navy, Westinghouse and Rolls Royce gas turbines to develop ECMs for other engines.

We even tried to anticipate new standard technologies by designing equipment based on the so called "Future-Bus" electronics. In the end the market for this did not take off. The control market was happier to stay with commercial standard VME and Intel chips. As I may have mentioned elsewhere, we did occasionally choose the "wrong horse", especially as there was so much choice. In the end every time this happened, the market always seemed to force us to end up on the most common commercial or consumer electronics. These were always cheaper, but usually just as good as some of the special invented control electronics standards. The same thing happened when we tried to use LANs based on standards other than Ethernet ie. FDDI and GM's manufacturing standard bus etc. The forces of commercial pressure always made Ethernet improve so that it was always as good or better than any more specialised standards. During this period even the US government began giving up on their military standard computers as commercial stuff was nearly as reliable and of course much faster and cheaper. Systems like our IPMS were the earliest military systems to use mostly commercial IC chips and many commercial standards. In much later systems of course we ended up using WINTEL PC computers too for much the same reason on all our types of systems.

Speaking personally, I think it is a good thing that commercial companies just go ahead and use what is on the market, rather than wait years for IEEE type standards based equipment. There are so many actors who try to muscle into the standards committees, that discussions go round and round in circles and by the time agreement is reached, the design has been overtaken by events and is obsolete. Even relatively small market segments such as SCADA took years to try and standardize SCADA RTU protocols. I have been in many such international meetings and it is a hard thing to keep awake during these boring discussions!

Eventually we had many naval contracts from the UK, Germany, Netherlands, Malaysia, Korea and so on. After I had left CAE, I noticed they have continued to win the newest designs for UK aircraft carriers,nuclear submarines and US nuclear aircraft carriers. I expect this marine business is still among the largest naval control systems businesses in the world. They have even built marine training simulators for submarines and have created a life time support facility for the UK submarine service.

The Sky

As a contrast with all our lucky breaks, a good example of the patience that is needed to develop a good systems business is seen in the long and tedious time taken to get our second ATC (Air Traffic Control) system. We won our first major ATC system contract in 1973/1974. The next project that we won was in about 1990. In the interim, we had many failed bids and fought off internal company attempts to kill off this business. Most of our other new control and simulator businesses came up either by luck or timing, rather than through careful analysis and planning. As was usual, the local Canadian connection was crucial in getting started. The main reason I actively championed an ATC business for CAE, was because I felt it was an almost perfect fit for CAE's real time customized computer control expertise together with the advantage of having developed one of the world's first distributed computer based ATC system networks. This gave us the all important and crucial experience opening allowing us to be able to bid other projects.

Canada has to manage one of the largest air spaces in the world. In addition it also provides access to the USA which is the busiest airspace in the world. There is a long tradition of flight in Canada, including the private use of small aircraft, as it is a huge country. So it was not surprising that the Canadian authorities (Transport Canada) specified a RFP for the most advanced computer based system for ATC in the early nineteen seventies. This was for the Joint Enroute and Terminal system (JETS) whereby the Enroute and Terminal areas in Canada would be put under computer controlled radar processing. CAE won this project together with an American partner team member who developed the new display generator system. This was an enormous software system with well over 200 computers arranged around 7 systems. At peak there must have been over 100 programmers working on this system. It was so big they had an entire building solely for their use. It was fortunate that CAE won this project as it coincided with an enormous drop in other business. In the end the project was a great technical success and operated essentially flawlessly for well over 20 years. However it was some years late and well over budget, so internally it had very bad press, and senior executive staff wanted to close down all future ATC business. After the initial development work had been done, I was given the job of finishing off the production systems and so had to take a lot of rude jibes about the losses whenever I was chatting with colleagues in the cafeteria. The only person who ever came to my defence was the company chief financial officer who reminded the cafeteria crowd that it was the cash flow from JETS that kept the company from closing its doors and going bust! Incidentally this also shows how easy going and friendly our senior executives could be, as they just joined us for lunch in the same cafeteria, ate the same food and easily joined in the usual bantering and discussions about politics etc.

ATC is mostly a ground based process for tracking various aircraft in flight within a region and then ensuring adequate separation (both vertical and horizontal) to ensure safety. At the time when we developed JETS the tracking over continental North America was by monitoring through ground based Radars located near major airports. Aircraft were expected to fly on special air routes along long distance radio beams that were sent from radio transmitters located on the ground. There were areas in the north and over the oceans where we had no available radars or beams. So in these areas the only tracking was done through long distance voice radio communications between aircraft and controller on special frequencies.

Essentially commercial air traffic is analogous to a stream of boats on a flowing river managed from the river side and kept apart according to safety rules. Except that the flow is in the air and in three dimensions and at high speeds. A further complication is the presence of other smaller aircraft that might be flying free flight at lower altitudes and they also have to be kept separate. Each region of Canada is divided for ATC into a set of control centers. Some are for Terminal control ie. control into and out of airports and others are for Enroute control ie. control between airports. Within each center the 3-dimensional sections of sky are further divided into sectors, each under a separate controller. As aircraft cross sectors there is a specially designed administrative process on the ground that hands over control of aircraft between sector controllers. In principle it is all commonsense, but can be complicated by the large amount of traffic and the interplay between individual pilots, not forgetting the effect that changing weather will have on flight safety. So an ATC controller has an extremely difficult job and the training may take many years. The Canadian ATC authority has a special college with a full set of actual equipment for the training. I visited this college many times for discussions on various systems.

The Radar technology locates an aircraft by bouncing the Radar signal off the flying aircraft and uses calculations of azimuth and distance to locate where the reflected system is coming from. Between Radar rotating scans the aircraft moves, so to get an idea of the aircraft track, it is necessary to extrapolate between return signals. A special class of algorithms called Radar data tracking algorithms are run on the ground based computers to produce forecasts of the aircraft track and this can then be visualised by the controller to see where the aircraft is going to, and what other traffic might get in the way. This is called radar data processing and tracking and is highly mathematical, as aircraft don't necessarily fly in straight lines. When multiple radars are available the best estimates of position are calculated by another mathematical algorithm called "mosaicing". Most of the time we at CAE only had one expert in this subject and were fortunate that he stayed with us all along. During long periods when we had no ATC work I persuaded him to help me with some of the marine work to retain his interest and keep him in the company. Without him we would have never been able to continue with ATC. It just goes to show how a manager has to keep key staff contented and interested during slow periods. It helped that at CAE there always was something interesting going on to keep our best engineers occupied and productive.

There are two types of radar signals. In one type the signals are just raw electronic signals bouncing back after reflection. This is called a primary radar and gives no information about what is being tracked. The controller would have to use other external information to correlate what is out there. The other signal is from a so called secondary radar which is actually an encoded message in the radio signal from the aircraft. This message gives identification and location information about the aircraft automatically.

The other method for monitoring aircraft was an accounting type procedure called flight plan processing and tracking. In this approach, using voice signals from the pilot (who tells the controller where he is located), the controller keeps track of his movements by writing down into computer files the flight's progress. This is then automatically printed out on special flight strip printers, with each flight on a separate strip. When flights progress, new strips are printed out automatically. Then these strips are laid out in order on a tray so that the controller can keep track of where the flights are in the correct order. When the aircraft eventually gets into Radar range the controller has to coordinate the paper strips with the correct radar signal to keep track of the flight. The controllers like this paper based strip system as it is flexible and they can even write stuff directly on the strips. Of course the ATC controller sends instructions to the pilot over pre-allocated radio frequencies. Hence for a large region the control computer needed a suitable large and fast database type system to hold and manipulate all the tracks and flight plans. The key tool for the controller to visualise the continuous flow of traffic in his sector was a very large map of the airways onto which each flight was positioned. As the map was two dimensional, but the traffic was flying in 3 dimensional space the controller had to develop a key skill in mentally converting the map in his head to a 3 dimensional picture. This led to the development of very large CRTs on which computers could position the aircraft. To keep the required accuracy for controller visualisation, the graphics had to be of extremely high resolution and accuracy implemented in vector graphics, not raster graphics, as at that time it was the only technology capable of such a high resolution.

Our American project partner had developed such a very high resolution (2K*2K) display generator and screen, which we further modified to integrate both the analog radar returns from the primary radars and the message returns from the secondary radars onto the same screen, together with calculated tracks. This was a unique facility in the ATC world for a long time. The reason why such a facility was required was because the client did not want to go straight to observing only synthetic signals, preferring to also have the actual radar analog bounced signals as a security feature in case of some unforeseeable error. The understanding was that the controller would switch between the cluttered but actual analog scanned display and the calculated synthetic display just to be sure. Eventually the synthetic display became so good and reliable, the controllers just gave up using the analog display completely.

As this was a prestigious project, the client insisted that our hardware design look good and be of the highest quality. There were specially designed sector suite control desks for the ATC controller that included keyboards, radio panel, 2K vector display, raster airways CRT and flight strip rack. A large control center like Toronto Enroute had many such sector control desks. All the many mini computers were rehoused in special new uniform computer cabinets together with the special CAE electronics for intercommunication and displays. The computers used were Interdata model 70, which at the time were state of the art. Each computer network was connected around a triple redundant LAN, specially designed by us. This was probably the very first LAN based distributed computer real time control system anywhere in the world. For each center there was a redundant computer pair for the main computer complex and its associated separate computer pair for just doing the Radar tracking. Each 2K sector display was driven by its own separate computer, also on the LANs. In those days there was nothing like a graphics work station, so we had to essentially develop our own mini computer based graphic work station for each display. To edit and compile the system and other off line work there was another associated computer system with the usual peripherals. This adaptation system was also used to tailor the system to the actual area that system was going to monitor. This system then could download new software packages or load modules to the online system as everything was networked. All data was recorded so that one could analyse an event after the fact in case of accidents etc. Every piece of hardware had extensive online and offline diagnostics. The programming language was still in Assembler in those days

After a very long factory system test for the first delivery system, each system had to be tested in turn for several months to verify proper operation. Then the systems were taken to the various sites and installed, followed by another long commissioning acceptance test. Over the years the client themselves set up a software team to modify and maintain the system over the years. At times they gave us some smaller support contracts for new modifications to the system. This system was incredibly reliable with achieved reliability numbers around 99.99%.

In those days system maintenance was a major cost and we had to train a very large number of the client's maintenance staff. This training was down to being able to replace components on PCBs. Each such system had to have a full set of spares including component spares, all of which made it quite expensive. Actually this was fairly common throughout all our control systems contracts, at least until modern technology became even more modular and individual spares were only needed for large separate modules.

Following this project, I tried to get a new ATC contracts from the Canadian military (called TRACS), but with no success. However we had an opportunity to bid for the full ATC systems in Yugoslavia and also for Greece,using the same team. The Yugoslav system came closest to being won, but in the end the politics were too difficult. These system bids dragged on for several years but there was very little executive level interest for ATC at CAE, so I had a difficult time keeping the idea alive. Occasionally an opportunity would arise for smaller systems at Gander oceanic center and at the Transport Canada R&D center, but it was only useful as an excuse for me to keep talking to Transport Canada engineers. It was on one such occasion, that I found out about their long term technology plan. They had produced a long term technology plan for ATC in Canada , showing how they would develop all their systems over the next 20 years or so. The plan was in book form called the "blue" book and was updated every year while they waited for government money. This gave me a clear marketing target to follow, while I waited for the next appropriate system. The systems that I kept my eyes on were a new radar system (called RAMP), a new flight planning and air traffic management system (called FDMP), a new system wide remote maintenance management system for all the Transport Canada electronic equipment, especially radio navigation aids (called AIMS) and finally a new ATC experimental simulator (called CAMSIM). Each of these programs were priced in the hundreds of millions of dollars and therefore were very tempting, for both the technical and business possibilities. As these systems were so large, inevitably the politics of Canadian provinces and the federal government would come into play, so I just waited to see what would happen.

In the meantime the defence department had purchased some long distance American Radars for monitoring the Canadian arctic against possible incoming Russian missiles. This was the North Warning system. The only possible Canadian contract was for central monitoring of computer communication systems which transferred the Radar signals to military control centers. This prime contract was given to a BC telecommunications company, and I bid for the communication remote diagnostic system using our military DATAPATH M electronics, but it was too expensive and they chose a normal telecommunication supplier's standard system instead. In those days telecommunications based computer systems seemed to be a possible candidate for us to diversify into, but the market was sown up by the main standard telecommunications suppliers and there was very little scope for a custom systems supplier like us.

The next Transport Canada project was an unusual one for us. The RAMP (Radar Modernisation Program) was to supply multiple new design Radars right across Canada. There were three competing team, two from America and one from France. To get the required Canadian content for the proposal, the French team took us on as a subcontractor to design and build remote diagnostic and maintenance software systems for the new Radars that they would be designing in France. So I spent some time at their offices and factories in Paris and Bordeaux getting the design details of their Radars to be able to design the remote diagnostics. It was very interesting finding out about high power microwave electronics in that new design. I admired the high standard of living in France. Even in the cafeteria, they served wine and superb restaurant level food. The French company lost, as they did not understand that they probably never had a chance against the Americans.

As these new radars would be initially connected to our JETS system, Transport Canada began talking to us about doing upgrades to that system. In the end they decided to just buy another completely new set of RAMP Radar display monitoring systems (DSE-Display Site Equipment), while connecting to JETS for all the other flight planning systems work. We were one of only two suppliers allowed to bid, the other one was the winner of the actual radars. As this was probably the largest bid we had done for a very long time, I had to set up a full engineering team to be able to respond. The bid was so large, our president himself led some of the sales effort, but in the end the radar supplier had a much better case as they supplied the radars as well. One of the problems with our bid was that by now our original 2K design for the display graphics was getting a little obsolete, though we did update the computer network (Ethernet etc) and offer high level programming languages. As this was an enormous bid the client gave us a detailed review of how and why they had selected our competitor. They claimed that they could have contracted with either of us, but that our competitor had a better price. It is possible that was the case, but I noticed that both the Transport Canada project manager and his ATC manager who had evaluated the bids resigned and joined this competitor one year after awarding that bid!

Many years later, another Canadian project management company came to see us about a bid to the UAE (United Arab Emirates) for a full turnkey ATC system based on the most modern equipment. They had selected an advanced American defence company BDM to supply the radar data processing, with us (CAE) supplying the rest of the system including hardware and flight plan processing. By now a modern full function ATC control system was based on OPEN standards. The computers were linked on dual Ethernet LANs rather than proprietary standard LANs. Each work station computer was used to manage the new design 2K*2K type of CRT. At the time only the Sony company could provide such a powerful raster CRT as they were developing the very first commercial HD TV type monitors. Until then only vector graphics would have been good enough. We were still using DEC computers, but had planned on converting our software to the C language with IBM RISC computers and UNIX. We still kept the flight planning computer pair and the radar data processing computer pair separate. So conceptually it was the same distributed architecture as used on JETS, including the links to external systems, but with more modern technology. Being a turnkey system, all the ancillary equipment such as inverter power supplies and civil works were to be included.

I had to make trips to the Arabian Gulf to meet the client and talk about the design with their British consultant. In those days a foreigner would only be allowed to land in a Gulf country if he was sponsored by a local company. So I made arrangements with a potential company to meet me at the airport. However when I landed in Abu Dhabi, I did not see anyone there so I had to sit on one side of the glass wall hoping to spot my agent. It was only after waiting for hours, I noticed a tired looking Arab gentleman sitting on the other side of the wall, patiently waiting as well. So finally we realised we were both waiting for each other!

As I was in the area I also went to neighbouring Muscat in Oman. They were planning a similar system. There was a great deal of competitiveness in the Gulf. It made no sense to me to plan two such very powerful systems within 10 minutes flying time of each other! During the final bid evaluation we all met at BDM's offices in Boulder Colorado. BDM's radar processing was extremely advanced as they had been selected by the US air force for their high desert training range. Tracking radar signals was very difficult when following fast moving air force fighters as they made high speed and sharp turns. They also showed me one of their inventions. It was an early design for cockpit displays that could automatically show a pilot all the air traffic around his airplane in real time. At the time the US FAA (Federal Aviation Authority) administrator was also visiting to check up on his own systems, so he put in a good word for us too. It was the first time I noticed how very hierarchy conscious the Americans could be in their defence industrial complex. They all seemed to bend over backwards to bow and scrape up to the FAA administrator. This was unlike us Canadians, who were used to treating even the most senior people just like colleagues. In the end the contract was won by a British company though we went twice through the bid process over a couple of years.

Probably around this time Transport Canada were ready to plan their FDMP (Flight Data Management Program) program, which they eventually renamed CAATS (Canadian Automated Air Traffic System). Their internal champion was the manager of their R&D facility who I had got to know quite well. He had planned the most modern system that anyone could have thought of, and occasionally called me over to discuss technical possibilities. As this would have been an absolutely enormous project our management wanted to bid the system as a partnership with another company so that the risk could be shared. For some reason that I cannot remember we decided that we would use IBM computers rather than DEC computers. I think it might have been because IBM had just come out with a new RISC based design for the IBM power series of computers. The computer companies had moved back to RISC computers as they had discovered that they could be significantly faster than the CISC computers such as VAX. Another possible advantage that we thought of was that IBM would have more political clout in Canada and so bring some much needed political strength to our bid. I think IBM may also have been a good fit because they themselves had an ATC subsidiary company. This subsidiary was at that time one of two providers of a billion dollar prototype system for the FAA. In fact this is where we got into some unnecessary complication. IBM thought that we were teaming with them for their FAA ATC software skills, while we were mostly interested in their computers and political strengths. In any case our management decided that IBM would be the prime as they seemed to be fed up with supporting my single-handed ATC forays.

So we began getting to know IBM a bit better. Their mind set was quite different to our way of thinking. There were actually two parts of IBM involved, the ATC part and the Canadian computer part. As this was a Canadian contract the lead was passed to the Canadian subsidiary which annoyed their ATC subsidiary as they believed that they were the experts and should have been in charge. So over the next couple of years I had a ringside seat watching how a giant company works. The IBM ATC company was a typical US defence contractor type company and started to overwhelm the project by allocating a huge team of over 30 staff to engineer and market the proposal system. I attended several meetings at their facilities on the outskirts of Washington and was quite appalled at their wasteful ways. If they had kept up this effort they would have used up all the potential margin of the system on just the bid costs. In fact I had to warn them through their Canadian staff to stop throwing money at the bid and to wait until the final RFP actually came out. The IBM Canadian staff was so overwhelmed by the IBM ATC staff, they actually preferred to deal with me as their advisor in what they should do. Computer companies generally do not understand the type of rough and ready project bidding that we were used to.

I must say though that all the IBM staff were very kind and loved to help. We were even invited to bid for their PAMRI updates to their existing computer equipment already in operations at the FAA. They were quite put out the first time they had a tour of our factory and saw the large number of DEC VAX computers that we were using, with not a single IBM computer in sight. Eventually the flight simulator departments did move their design to IBM and in ATC we also started using IBM machines for all our bids. They brought a lot of strong marketing talent as they had a good insight into the Canadian federal government procurement politics. We had several meetings with their lobbyists in Ottawa getting market intelligence about potential competition etc.

In terms of actual experience in producing such a huge ATC system the best American candidates were IBM and Hughes. The reason was that they were both contracted to do a design for the planned American replacement systems. The budget for these American systems was in the order of $25 billion! So they had let preliminary design contracts to both competitors and would chose one after a long 2 year design effort. So both IBM and Hughes were planning to bid the Canadian system as a fall back in case they lost the US system. In the end IBM won the US system, so Hughes made a big push on the Canadian opportunity. They got the politics right by offering to set up a brand new facility and Canadian company in Vancouver BC where they would build the Canadian system. They also offered DEC/HP computers. So essentially the political fight was between us based in Eastern Canada in Quebec and Ontario and the opposition based in western Canada in BC. At the time the transport minister was from the west and it was felt that it was the west's turn so they won, even though the actual project team seemed to prefer us, as I found out from my friend the Transport Canada Project manager. This Canadian government method of sharing out work on the basis of regional developments really does not work well. By constantly creating new regional companies through subsidies, they actually waste the money as rarely do these artificially created companies have the staying power to continue in business after they have finished the initial contract. So sure enough after Hughes finished their contract they closed up shop in BC.

However from a technical point of view designing the system for the bid was very interesting. The client was as I have mentioned, very technically shrewd. They wanted essentially a system based on all the most modern technology and based on all OPEN standards. There were the usual 7 regional systems that would eventually replace our old JETS system. They added the interconnection of systems such that nearby control regions would back up the nearest other systems. This meant that the computer networks were very complex and huge amounts of data were to be transferred between systems in real-time, which required very powerful computers. As the supply was turnkey all the support systems and power had to be supplied as well. On the application side Transport Canada themselves had specified a large number of applications with detailed instructions for each ATC controller task that we would have to program into the system. They had specified a huge amount of spare capability so that these systems would easily last many decades. One forward thinking item was the requirement for the software to be easily ported to more modern computers when existing computers became obsolete. They had also given up on their usual requirement to be trained to repair equipment in detail as they had usually specified in the past. When I was surprised by that, their project manager asked me when I had last had an old TV set repaired. It was just not worth the expensive labour and training to repair modern electronics. It was easier and cheaper to just replace the boxes. This was quite unusual for me as most of my previous clients had always insisted on component level repair capability. After this episode I "modernized" my thinking too. This was probably the most expensive bid we had ever done and the whole effort from start to finish took nearly 3 years.

Our luck finally turned after this big loss. My old friend the Transport Canada project manager also happened to be an important technical member of ICAO (International Civil Aviation Organization). He was helping the Icelandic ATC authority with their plans to update their equipment with a brand new ATC system based in Reykjavik. US, Iceland,Canada, UK and Portugal controlled the air space across the North Atlantic ocean. Each country had its own area of responsibility before passing on aircraft to the next authority. Canada was the first to invent a new type of computer control facility called an Oceanic control facility based in Gander Newfoundland. I had learnt how this worked while looking for new business. So our friend suggested that we try and win such a similar system for Iceland too, but of course with modern computer technology. Oceanic control can only be done by using flight extrapolation algorithms run in the ground based computers, based on radio messages from the pilots. This then allowed the ATC controllers to keep adequate separation between aircraft over the North Atlantic. They monitor aircraft using special conflict probe calculations which make sure that multiple aircraft do not cross each other very closely. There were obviously no possible ground based Radars for flights over the oceans, so Oceanic ATC required a quite unique new technology. We won this contract and as a consequence we became the sole experienced Oceanic ATC supplier world wide. This system had the usual flight plan processing and management software but this time the flight strips would be replaced by electronic flight strips on the controller CRTs. Once again we were in the forefront of technology innovation, as this system was probably the first ATC system in the world to use electronic flight strips. We learned to implement the usual conflict detection algorithms and techniques of how flight space was reserved, allocated and released. The system had to work with weather reports from the UK meteorological office and chose the correct north Atlantic track system depending on the winds aloft. We also trained a local Icelandic company to maintain their system locally. A consequence of this win was that we also won similar systems for Portugal, New Zealand, and Hong Kong, as all those areas were also surrounded by oceans. Eventually even the US FAA through their ATC prime gave this part of the company similar contracts for US Oceanic airspace. By then however this business section had been sold off by CAE.

In the meantime I had started to plan our attack on the next big transport Canada ATC system ie. the AIMS (Air transport facility Integrated Maintenance System). At the time we felt (because of our SCADA experience) that we had to be the favourite to win and other companies came to see us to try and join our team. This system would have automated the remote diagnostics and maintenance of all Transport Canada electronic equipment out in the field all across the country, especially the radio navigation aids. So I spent a great deal of time at the Transport Canada technical facility understanding what modifications and interfacing details would have been required. Unfortunately by then government money was in short supply and they cancelled their plan for a country wide system and replaced it with a very small trial system which was too small for us to win.

Eurocontrol came out with a very large bid to replace systems in Germany,Belgium and Holland. So I went to Europe and tried to arrange a team with our CAE subsidiary German company in Aachen/Stolberg. We had meeting with Eurocontrol staff in Brussels and were pre-qualified to bid as were only 4 other companies (Two American and 2 European). Unfortunately it was not to be, as our executives stopped the bid just as I had finished most of the initial design. Probably it was a wise decision as the Europeans were not going to allow non-European companies to win. This is exactly what happened as the project was divided up between the two European companies. I did however get to learn something about European ATC software standards and flight control procedures.

The very last ATC system that we won before the ATC business was sold by CAE, was by far the most complex ATC type software system in the world. I was very keen on this system as I could see a huge amount of development work that the winner would get and so would then have the most advanced technology for future use. This was the Transport Canada CAMSIM (Canadian Airspace Management Simulator). This was such a complex system that I do not believe any company other than CAE could have done it single-handedly. In fact one of my ex-colleagues who had by then left to work for Hughes/Raytheon told me that it was too risky for their management to take on.

As I have mentioned, Transport Canada was a very advanced customer. To manage and plan their ATC procedures they had a full ATC research simulator which they wanted to replace with a much more advanced simulator. The computer network we chose was a redundant token ring FDDI computer network with a very large number of IBM RISC computers networked on this system. In the airspace simulation computers, we had to simulate an absolutely enormous airspace the size of pretty well the whole of Canadian airspace, with all the usual ground based systems set up at airports and along airways. The weather had to be simulated as well as a large number of detailed aircraft flight simulations that could be run to represent traffic in the airspace. There was full range of choices of different aircraft types that could be selected from. Essentially we had to use all our existing skills in flight simulation to produce this huge number of accurately modelled aircraft. There would be a number of pilot stations from which pilot station operators would actually fly about 15 aircraft per station, depending on the planned test. In parallel there were a set of development ATC consoles on which the test ATC operators would try out the test scenario. These included specially developed full graphic 2K raster CRTs which could be modified to test various situations and various graphic software types. A subset of all the existing actual ATC control stations (both RAMP and CAATS) were also networked into this test equipment. There was a management facility and Instructor station to plan, run and finally record and analyse each trial test. The system had to include a full suite of normal ATC software such as had been required for CAATS, so it was also our first new fully integrated ATC software package for both Radar and flight plan processing, including oceanic processing. Such an experimental simulator also has the facility for " Fast Time Simulation" which allows the team to rapidly pass over the test exercise to get to the situation that needed the detailed real time simulation.

Finally there was a parallel voice communications system to mimic the pilot-controller traffic. For this we had developed a full all digital computer and software based communications FDDI network together with pilot and controller communications panels and electronics. This was called SIMCOM (Simulated Communications). We had planned to use this new development as the basis of a fast all digital new communication system which we could market to ATC and other control centers. In fact we took the risk of all this graphic, networking and communications new developments, thinking that it would give us a well established new technology base from which we could actively sell other modern systems. Unfortunately it was not to be, as there was a change of senior executive management who lost their nerve and sold this business after we had done all this development. The lucky purchaser company now has all the benefits of our development work.

12. A short discussion on Technical and business issues.

One of the most worrying issues that has resulted from the highly networked computer systems of today is the issue of contamination. This can be inadvertent, in which case it is essentially just another bug, though much more difficult to evaluate. The worse issue is deliberate and malicious faults put in by hackers and criminals. I suspect there is not too much one can do about this second problem other than good old fashioned security practices such as doors and locks to physically isolate equipment. When working with nuclear systems I noticed that safety systems were sensibly kept so simple that there was little chance of faulty logic. In addition it was all done only in hardware so there was no chance of errors arising from software issues such as those from compiler errors.

One approach that could be used is to design computers with safe areas with read only memory for program code. To allow write access to this memory one could design a manual switch that could be used to make this memory temporarily write access. We did this in the past on several occasions. On drum memories we had each track protected by a switch. When it came to memories on our intelligent RTUs we used special EEPROM (Electrically Erasable PROM) memories. There are many memory technologies available that could be used to protect programs or make it only possible to access memory under very well protected procedures.

There is a bigger problem when it comes to protecting data, especially data that is used by the operating system to keep managing the computer systems. Too many designs work with flexibility in mind rather than compromising and looking more carefully at security. One approach from the past that is easy to replicate is to have a quick reload-reset facility. This physically restarts the entire system from code that is in write-protected memory and thus uncontaminated by corrupted link data. Almost all our control systems had such a restart facility, as after restart the system could work out where the system was at and then carry on. In fact in many cases we had a core memory mirror already set up on the disk so that a restart was a straight forward down-load directly into memory followed by program run. Corruption of the long term data base can only be protected by careful backup procedures and keeping multiple copies. All this was normal procedure for control systems and I cannot see why the same thing cannot be done by private commercial/home users. In short what I am saying is design things so that it is easy and fast to restart the whole system from scratch including system management data and put long term business type data into a physically separate device with a protection switch.

In the past we designed our fail safe and redundant systems by including parallel duplicate systems with a fast changeover, This gave us very high availability and reliability figures eg. 99.995% in the case of our ATC system. The element that we did not foresee was outside malicious behaviour. Another flaw in some of our designs was when we used identical software running in both main and stand by computers, if there had been a software bug it would happen on both systems at the time of changeover. The best way out of this problem is to have some differences in the timing or sequence of software in the two computers. In the most extreme cases one could have entirely different software written by two independent teams, but it would be outrageously expensive to do this. Only the most important systems could justify such an approach. I never came across such an important system in normal industrial systems, so a compromise design with sensible restart-from-scratch type features is usually more than enough. After all even battery backup will eventually fail if mains power is down for a long time. We usually backed up our RTU memories with batteries for just this reason when the technology of RAM replaced magnetic core memories. In addition we always included a hardware remote reset/restart facility on each RTU, After all RTUs are far away in the field and it was impractical to send staff out to restart equipment on power fails or errors.

Computer network architecture has progressed over the years from a hardware hierarchical design to a hardware network type design. It all happened in stages, as new technologies were invented. Since we made it a point to always use the very latest equipment, one gets a good overview of the progress just by looking at our systems architecture chronologically. For instance in the James Bay and Nuclear DCC systems each computer had its own (AYDIN) display generator and its own front-end, so it was only possible to changeover the entire system when any one device failed. On the GURI and Hydro Quebec regional systems, the VAX computers were connected to the AYDIN display generators through new dual interface cards , one link to each computer from each subsystem. Similarly each DATAPATH front-end chassis also had dual links, one to each computer. So then one could just changeover a failed subsystem rather than the entire system. When we designed LAN networks for the New Jersey system the whole hierarchy became virtual as the LAN design did not constrict which subsystems worked with which computer. In this design we could even re-allocate functions to multiple computers. In GURI we had an early version of this type of re-allocation when we used computer shared memory to act almost like a very fast LAN. Once we started using individual graphic workstations directly on the LANs instead of shared chassis graphic display generators, the system became completely flexible. Even secondary devices such as printer servers and network gateways to external systems became independent computers just connected to the LANs. It even allowed maintenance to be carried out on equipment while other equipment was actually doing the control in parallel. Training consoles on the marine and power control systems of this era can work independently with simulation computers without disturbing the actual control system computers. One can even design a hierarchy of different physical LANs for various parts of the network. In this manner one can physically isolate crucial LAN sections for security and only do inter LAN connections through isolatable gateways.

One of the peculiarities of software design is that stuff reverts to old ideas time and time again. In a sense there is very little that is completely new in software and systems. For instance recent PC software designs favour keeping data in the cloud. This is very much like the old days of time sharing central computers and limited function VDUs (eg VT100). This cloud approach helps users by passing all the inherent data managing and security issues back to centralised experts. In real-time industrial type control systems, we see the same effect too, whereby the fashion moves from centralised systems to distributed systems and back to centralised systems. Often this effect is driven by the shortage of skilled staff when too many distributed systems are used. In theory distributed systems can make sense, but organisations have coordinating difficulties if there are too many distributed systems to manage. In the past we often designed distributed computing networks in control centers, simply because several smaller computers cost less than one large computer. In such an approach, we allocated software among the various computers at one center depending on how much algorithmic calculation had to be completed in parallel.

It is probable that very large calculations will continue to be handled by computing sub-sets of the calculation in parallel machines. As long as that computer network is loosely coupled over standard links, the complication effects are quite manageable. The programming problem of truly parallel tightly linked multiprocessors is too great to be of much use in general, though it could be of some use in certain highly compute intensive industries like weather mapping.

An interesting element in software design is to compare how different data structure designs are chosen for different applications. The first time that I noticed this was when I began to analyse ATC (Air Traffic Control) systems. In such systems there was a move away from a shared central fixed data structure area (essentially a data base area) from which the various programs could get the raw data on which they could work. In our design for the flight simulator, nuclear simulator and control systems we used just such a shared area data structure from which individual programs drew their working data. The designers of our ATC system based the core of their design on a data structure approach which mimicked the flight flow. So the data effectively flowed through the system as the actual aircraft flowed above. So it became a message passing type of system. I do not know why this approach was chosen, but it was conceptually easier to debug as data tracked the actual aircraft progress. I think other transport type systems also use a message passing type of core design. It goes to show that all systems are very closely linked to the administrative type of manual process that is being automated, unless there is an arbitrary historical reason for the design. Shared area central data was a natural choice for continuous type monitoring systems, while message passing type systems appeared to be natural for episodic flow systems. I assume other systems such as ticket booking systems and bank transaction systems are also episodic in nature and probably also work with message passing systems. One of the design techniques used by very secure operating systems and secure programming languages is to isolate individual software processes completely from each other and only pass on data through a secure message passing system, this is what the old GEC 4080 that I used in England did. That way there is no central data base area to be corrupted when something goes wrong

One of the ideas that I favour is the use of high speed accurate computer simulation to track industrial processes in parallel with measuring the actual process parameters. This can then be used to evaluate when the actual process starts to drift away from the desired theoretical simulation provided process. If the two differ by a significant amount an operator can be warned to take corrective action. This same approach can be used to find out what is causing a problem in an industrial process. The parallel fast simulation can be run on several scenarios starting from different potential error points. The simulation that comes closest to the measured effects could then inform the operator of the likeliest cause. As industrial systems become more automated, the operator has less knowledge about what the automation is doing within the actual process, so a fast "Tracking" simulation with an appropriate HMI can be used to keep the operator up to date on what is happening.

Some elementary versions of this approach were used by us in doing security analysis and state estimation for power grid security in EMS centers. However I could not find a client for more sophisticated real-time tracking simulation rather than planning or after-the-fact simulation. At one stage we did get some R&D funds to study the effect of using Artificial Intelligence (AI) to evaluate problems after simulating processes. This project was done in partnership with OH (who supplied the target industrial process, ie. Feedwater pump arrangement in a nuclear plant), University of Toronto (who supplied the AI expertise) and us CAE (to build an actual prototype system). It was a technical success. In one case it even showed that the existing automatic normal control strategy was faulty, even though it had been in use for many years at the power plant. The arrangement of pumps and flow valves was such that the control algorithm kept on opening the valves in an unbalanced way until valves on one side were fully open and could do no more, at which point the station could produce no more.The correct strategy would have been to open the valves in a balanced manner, if only one could have seen this effect. It was only noticed while using the tracking simulator to test what was happening. I am surprised that after this quite clear result, that OH did not implement such a tracking simulator for the whole plant. Maybe there was some other issue that I did not know about.

We should have left aside funds for some marketing effort so that we could have persuaded a real customer to try this approach on a larger plant sub system. Unfortunately the funds ran out and management dropped this effort. I expect sooner or later some company is going to do some more along these lines, as the problem of complication in supervising automated systems is not going to go away. Some years ago there was a crash of a new Airbus aircraft (which has computer control and fly-by-wire management). The probable cause was that the pilots could not understand, or were not following the logic of what the automatic controls were doing, especially as they were not in the loop. This meant that when they had to go to manual control in an emergency, they might have started to work against the automatic strategy that the automation was going through and so crashing the plane. The key difference this "tracking" approach would have versus the more common approaches that monitor individual pieces of equipment is that "tracking" looks at the entire plant as a whole and may pick up problems that do not show when looking closely at individual items. It is worth remembering that systems as a whole can behave quite differently from the expected behaviour when considering individual system component items. After all this is exactly the problem of debugging system wide software, after debugging individual programs.

Another type of industrial control support system is the increasing use of computer health and condition monitoring of rotating machinery through the use of sophisticated algorithms. We used these concepts in managing marine gas turbines on our naval machinery control systems. In addition to the calculations in the local microprocessors, this required a new set of PCBs that specialised in locally calculating variable parameters such as vibrations and pressure trends. To do this we used special DSP (Digital Signal Processing) ICs on the DATAPATH boards. These dsp chips allowed more sophisticated Fourier type maths to be done at a fast rate. I even investigated a potential bid for a prototype HUMS (Health and Usage Monitoring) system on a helicopter. As this was a new concept the helicopter manufacturer wanted the potential supplier to also take some of the development and cost risk. Our management were reluctant to get into the business of supplying actual on-board electronic aviation devices, so they turned down this offer. It does show the type of complex systems that potential clients would send to us, indicating their confidence that we could potentially do such a project. In the power industry on the transmission side, health monitoring can be easier, essentially just keeping track of time in continuous use or time in over stressed use. We did use this concept on monitoring transformer insulation condition. As utilities have hundreds, or even thousands of transformers, keeping track of their operating condition is a very valuable project.

Advanced on-line maintenance support of machinery is becoming more common. On board diagnostics are getting more and more useful. Even in the days when we bid the RAMP RADAR systems, the client had specified sophisticated on-line diagnostics that monitored the condition of the RADAR electronics while it continued to work. The larger AIMS system that was planned by Transport Canada for all its remote navigation aids and radio beacons was also similarly an on-line continuous diagnostic facility. In fact the shortages of competent maintenance staff, plus the complication of modern electronic controls,demands equally exhaustive on-line maintenance aids.

 I did some work in designing a concept for such an Advanced Technology Support Center (ATSC) for specialist maintenance companies, while I was an independent engineering consultant, but the idea has not properly taken hold yet. Such sophisticated centers will be required sooner or later as systems become so complicated that maintenance of the systems will become an organisational bottleneck. It is another version of the original trend whereby automation helped reduce costs and brought accuracy to industrial processes but added complexity. I predict major growth in the requirement for systems companies to also take on many of the service and maintenance needs of their customers. This is being driven both by the complication of the automation and the shortage of experience at the customer end. So we have to help automate the increasingly complex task of maintaining these increasingly large and sophisticated complete systems. I notice the commercial airline industry have already set up on-board systems that remotely send data to maintenance ground stations while in flight, enabling them to have maintenance fixes ready on the ground for when the aircraft lands.

These maintenance devices rely on conventional diagnostic analysis. I am suggesting that with the advent of very fast computers it should be possible to invent diagnostics based on very fast tracking simulation to give even more accurate "system-wide" analysis of problems that show up only on a full system basis. This will supplement the current set of individual device specific monitoring systems.

The big unknown is the effect of exponential growth in computer power. Quite a lot of people are foreseeing the day (perhaps not so far away) when machines can do the work of humans etc etc. Maybe! However in the interim I can confidently predict there will be many systems built that will require humans to fix the inevitable errors that will occur. So in the interim (which could be many, many years long), one must have technology that can help humans fix complex system wide problems. After all I am speaking as a practical engineer not a scientist or futurologist. Even automatic checkout stations at modern supermarkets need a human clerk hovering nearby to help fix the inevitable customer or machine "snafus".

To summarize, simulation systems are required for the three Ts ie. Training, Testing, Tracking. So far we have well institutionalised simulation for Training and Testing, but only partially for Tracking!

Software development is still something of an art rather than a normal engineering discipline. A great deal of effort has been spent on making the actual coding easier and more repeatable. This is represented by the way we went from machine language programming through assembler languages to higher level procedural languages. We even developed some graphical process type languages where programming is abstracted behind a graphic icon and simple flow charts. One of the problems with this approach is that it only tackles the most measurable of the stages in software production, not the most expensive and difficult stages which are design,integration and test. This is a normal human approach where one only concentrates on what one can measure even if the benefit is marginal. If we want to see real as opposed to "political" progress we have to improve the design and integration stages. Until that happens, software production is a bit like a science experiment full of trial and error. A good project manager should accept this reality and work with it rather than pretend that things are going well when "coding complete" boxes are ticked!. Another issue with high level abstractions like objects or icons is that behind the icon one still has to do the difficult work of writing the actual code, which has not become much easier. So in principle the total time to develop icons and final programs is usually about the same as doing things the old way. Hence I am slightly sceptical about propaganda on modern programming techniques.

I have noticed that software productivity never seems to improve by much in the aggregate. Of course some individual programmers can be several times as fast and productive as the average. In general, work is done in a craft-like fashion, and managers have to rely on individual skills only. Tools do not seem to make that much difference, in spite of the propaganda. For instance years ago we used to have special typists who did all the initial coding preparation, on the theory that the typists were faster and less costly than programmers doing their own typing. However when individual CRT terminals became inexpensive, programmers insisted on doing their own typing and no amount of management pressure could get them to use the typists. They found that doing everything themselves was worth any extra cost. The story about other software tools is similar, they can make life a bit more convenient, but it never shows up in better costs or a faster schedule. It seems that only the basic tools are enough ie. editors, compilers and debuggers.

I remember one absurd instance about how mixed-up programmers can get. When we were developing the prototype marine control system, we had to produce only about 20 control operator display pictures for use by the HMI display generators. The software team automatically assumed that they would have to develop software for a new display compiler program. This compiler was only required once, and would never be used again. I suggested that as there were so few display pictures we might as well just hard code each one by hand, The team were horrified and insisted it would be faster to design a compiler and use that to quickly develop the pictures. Needless to say, it took several months of hard work to get the compiler working, and blow both cost and schedule. It is difficult to force common sense productivity onto programmers. In fact there is a historical story that the very first SCADA picture compilers were developed by programmers who were too lazy to build pictures the hard (and boring) way, irrespective of the cost to a single project.

The use of Software QA (SQA) has become a big issue for modern Real Time systems. There are rigid standards that have to be met by an organisation to certify it as one that produces reliable and repeatable software. Over the years we made it a point to make sure we were always certified to the current highest SQA standards. To some degree I can say this is done for marketing reasons and to reassure clients. Often it just adds a great deal of bureaucracy and cost and has relatively little advantage.

I am not entirely pessimistic about software manageability. The most likely method that will bring about truly reliable repeatable software is probably the use of well specified industry specific software objects, especially larger more complex objects. Modern PC Wintel type technology makes this more feasible. The reason why I prefer reasonably large objects is because small objects still lead to essentially just another type of high level language. So the software development is not going to be much different than normal high level language programming with the usual logic and sequence potential for failure. Complete large objects will have been tested many times in different systems and hence new combinations of large objects will need relatively little testing as it usually will only be a sequential arrangement of objects. If these views find favour, it will imply that a successful system design company will build a good set of relevant large objects that MUST be reused on different systems, in the same way that a mechanical company will use the same set of gears to make different types of gear boxes. This implies that a systems company must understand its business so well that it can define these generic objects. This is quite hard to do, as often one gets pushed into making smaller and smaller objects, and that of course is not a real solution. One reason why this is relatively easier now is that computers are so fast, one does not have to worry too much about efficiency or program size and so can concentrate more on generality. Of course this does not do much for basic systems software ie, drivers, operating systems, electronic interfaces, networks and so on. So far I see no solution in the future other than good experimentation type trial and error!!

One enormous advantage of the large numbers (millions) of users of PC type devices is that any software bugs are found quickly as the large number of users means that every eventuality is tried out. So general commercial public software becomes very reliable as far as finding design and coding errors is concerned. So maybe some use of such general public subprograms in real time industrial systems may actually make them more reliable. This is possibly a hopeful development. Already many computer related factors eg. the computers themselves, network hardware, operating systems etc have become commodity type items. This has probably reduced the amount of custom engineering required in developing new simulators or control systems. It should reduce schedules and cost. However there is still a tendency to increase functionality, so maybe there will still be a large amount of custom work required on these types of systems. It is unlikely that systems companies will be able to substantially reduce their systems engineering departments.

How to charge for services in a project environment is a tricky matter. Some lucky professions such as law can charge by the hour spent irrespective of the result. Most industrial /aerospace/ military systems contracts are fixed price, even if there is a lot of new development involved. So the risk is generally taken by the supplier not the client. This can be re-negotiated in some cases, when the technology development can be shown to be much more difficult than could have been foreseen, but that is rare. There are some advantages to fixed price bidding, as it forces the supplier to maintain a disciplined approach and to avoid wasting money. Some government contracts, especially in the USA are priced on a cost-plus basis. These contracts go on for ever and often do not deliver a workable result. I tend to prefer fixed price contracts as it avoids the problem of having the customer constantly looking over one's shoulder and interfering. In most cases if one has enough experience, one can get a fixed price bid to the correct price, and make a reasonable margin. Generally there are enough parts of a system that are well understood and these can make up for the unknowns. Getting a price right is a major skill that a good systems engineer has to develop.

One of the hidden advantages that we had at CAE and at GEC was that we became very close to large but local users and clients, such as the CEGB, British steel,Hydro Quebec, Ontario Hydro, Transport Canada, the Canadian Navy and Air force and so on. This is a vital element in the growth of a systems business. The only way to be allowed to bid on large and complex systems is to show a track record and satisfied customers. So in order to start, the very first system is the key. The first customer takes a big chance in choosing an unknown player. Not only did our local customers take a risk when they chose us, they also allowed us to develop the very largest and most complex systems in the world. This was because they themselves were also very talented engineers and users so they invariably specified the most advanced systems. When we had finished those systems, we overtook our competitors. Behind all this were the British, Canadian and Provincial governments who had used public funds to start these very large utilities and organisations. Unfortunately for a long time we did not have good access to the actual federal government bureaucracy of Canada, so we never got a reasonable crack at some of their other bigger real time software systems. I think it was because we did not play the political game. It was usually the lead engineers of the large utilities who favoured us and fought their internal battles to get us selected.

An interesting wrinkle that I noticed across all the system types that I worked on was similarity of the manner in which they had all extrapolated their original historical manual management and supervision approaches into computer based automation. In every case it seemed that it was essentially all based on a kind of detailed accounting and auditing just like the way finance departments have managed reserves ,balances, cash flows and kept accounts for centuries. For instance high level management of power systems and ships meant keeping track of energy and machinery condition and recording everything that happened just like a businesses' financial type accounts. In air traffic management it involved keeping accounts of actual traffic flow and making sure movement costs were optimised. Essentially in every case there is a human system (manual or computer based) that actually follows exactly what is happening in parallel in the actual physical world. It seems to be way we humans can trust and understand what is happening, rather than letting things rip and just looking at the final results. At the heart of every system was the collection of huge amounts of data for historical analysis. I guess managing any type of business however technical it may appear is based on following events with an accounting type approach. Even the training simulators were just automating the previous class room type of manual training methods and techniques, finishing up with reports and certificates just like a school. So application specific knowledge was only available from the actual original users who had eventually decided to put their manual procedures into computers. I only realised this strange commonality while I was helping implement the CAE ERP (Enterprise Resource Planning) system which appeared to do for our company engineering,manufacturing and financial data what I used to do for my control customers.

The only way that a small country such as Canada can build really advanced technical infrastructure and skills is by government subsidy. This is an issue that is well hidden in the usual "free market" propaganda that floats around the western world. In no country (large or small) has a significant initial technical capability been created except by government help and subsidy. After all governments do not hesitate to subsidize university education and research. So logically they ought to subsidize or give preferential treatment to advanced industrial work done in country. It is a form of learning that is much more useful than any university course. In my travels around the world, the most important element for many projects was the transfer of technology and training to help build local capability. There is a false idea that one can somehow hide one's technical crown jewels. It is better to help one's customers to become self sufficient and rely on your own ability to keep innovating. Usually electronic and computer technology develops so fast, that it is pointless trying to hoard technology older than a couple of years, . It just becomes obsolete and new clients will not want it. Unfortunately most political thinking is predicated on production type industrial business, so governments come up with techniques to protect intellectual property that are of little use in the systems business. During my entire long career we virtually never patented anything in spite of being the first to invent new systems, but we still kept up the tempo and built larger and larger businesses with reasonable margins.

13.Some Observations on Hardware-Software Organizations

In the beginning all engineering was finished as only hardware. Everything made, was made in some sort of hardware. So companies organized themselves to facilitate the making of concrete things. This was true in production type factory environments, but also in project type civil or mechanical engineering construction or building companies.

A typical factory consisted of departments or groups for the following operations activities: Design engineering, purchasing, manufacturing, testing, quality assurance, field installation and finally commissioning. They of course were supported by finance, sales and project management. One of the consequences of this way of working was that each department was only involved in a part of the creative activity. As things were compartmentalized, in practice no one knew enough about the whole product, so a "silo" mentality prevailed. To solve the most difficult problems a team had to be brought together to discuss the issue including members from each department. This worked well enough, until software entered the picture. In hardware, bad material or implementation will cause faults. In software it depends on design rather than implementation as bad implementation is usually found out during testing etc. So the issue of performance and reliability is much more subtle and harder to define.

The problem with software was such that it was extremely difficult for a programmer to debug or fix a program that had been written by some one else. Each human seemed to think of programs in a unique way! A consequence of this was that software programmers had to combine all the above hardware related activities into the same person from design to code to production, test, install and finally to commission! At least that was how things were organized in those early days. In later years attempts were made to divide the work into similar silos as on the hardware side. My own opinion is that it is mostly a bureaucratic façade, except when dealing with trivial issues. The sheer intangibility of software makes it very difficult for someone on the outside to get into the mind of the other programmer. I have done it on occasion, when some one has left a half finished job, but it takes forever. Often it is easier to scrap the program and rewrite it from scratch! So a good approach is to allocate any significant (suitably large) design task to only one designer so that everything is kept in just one mind. Otherwise defining the system and data across many designers will invariably end up with all sorts of assumptions being forgotten, leading to the corresponding errors. This way of working is called "chief programmer " teams and all software should be built this way. On very large projects one could have two layers of chief programmers, but generally this is not advisable due to the complication. If a system gets too complicated it is better to break it down into several virtually independent systems, even if that appears to be expensive and redundant.

There are several consequences of this condition that makes managing and controlling software an almost impossible task! There is very little tangible that a manager can see and touch to prove that something is ready and fit for purpose. So in those early days everything was managed by trust. Software teams were left alone by the management until the job was complete. Every attempt to monitor intermediate steps was mostly futile. An interesting consequence was that as a programmer one did a complete job from beginning to end including getting the program working in the field. We worked more like our medieval craftsmen forefathers than like modern industrial workers.

During my career we went from a total lack of any software project tracking to various techniques depending on the fashion of the day. The earliest established fashion was for the so called "water fall" diagram. In this approach the programming work was formally divided into stages (each one with an actual tangible output) ie. specification, general design, detailed design, coding, unit test, system integration , system test, and final system commissioning. At the time it seemed a reasonable approach, but it actually hid some of the most important elements of programming. The early divisions up to unit test are easy to list and record. So this gave the management "types" a feeling of being in control and knowing what had been done. Unfortunately the reality was that it proved very little. Most complex software has to be redone in several iterations as unusual flaws only surface when various elements are put together. In fact if sufficient faults are found, its safer to rewrite a program than to keep modifying it as the changes can clutter up understanding. So a new technique was invented called the spiral technique or iterative technique. This was good for writing theoretical studies, but it still did not provide any real intermediate tracking points. The only time one knew one was finished was right at the end. Until then everything was tentative. Hence many if not most of our development projects went late and sometimes over budget.

Another issue that often arises is the issue of hierarchy. Historically I believe the modern industrial corporation organised itself by using the military or the church as a template organisation. These organisations rely on a deep hierarchy of levels and managements to supervise and run the organisation, divided into "silo" like divisions. In such a scheme, decisions have to go up the hierarchical chain and then down a parallel chain to get to the actual workers. This type of organisation is hopeless when it comes to doing custom or newish work. Large static bureaucracies can be very good at making well understood and standard products, but only add costs in an innovative environment. So the correct organisation is some kind of network with very little hierarchy or many levels of supervision. In such cases the actual worker is usually the best person to decide. The boss mostly should keep out of the way, his primary job becomes one of coordination and support rather than decision making. Incidentally even the military actually changes its organization when actually at war. Most battles are fought "project " style by gathering all the silos under a single general outside the usual hierarchy.

One related issue that is worth a further comment is that of the best relationship between line and project management. I have worked in three major companies ie. GEC, CAE and SNC-Lavalin. Each one had a different approach. As SNC was an EPCM (Engineering Procurement Construction and Management) company , the project manager was a "king". Every one working on a project reported to the project manager who was essentially like the "president" of his project. At GEC the project manager was in charge of the project to the extent that he controlled and was directly responsible for the customer interface, the budget and technical choices. However he was less involved in the detailed work which was the responsibility of various line managers. CAE was the strangest of the three. The project manager had no authority whatsoever! He was only entitled to act as a customer interface and report what was happening to line managers who actually ran everything. My personal preference is for the GEC approach where the project manager has authority, but is still part of the bigger company network. The SNC approach is really only useful for construction type companies where a single project may be worth $1 billion, such as a project to build a complete power station. The CAE approach leads to very frustrated project managers and relies on the very powerful line managers getting along with each other.

Another issue, especially for software managers in expensive labour countries, is the issue of outsourcing. Salary levels in such countries can be several times those in countries with low labour costs. As average talent can be similar in either country, how can one justify doing software in an expensive country? I think the answer depends on whether advantage is gained in being very close to the customer and in the advantage of a tightly knit team. The team advantage is crucial. Just as it makes sense to keep the same programmer on an entire job from beginning to end, so it makes sense to keep teams small and close knit. This will avoid problems of design or data misunderstanding that will occur between programmers. The errors that are likely to occur with long distance teams are so great, that it can wipe out any cost advantage of working with cheaper labour. The more complex the project, the more important it is to have a smaller team closely co-located. The final issue is the great advantage certain talented individuals bring. It is easy to see that some programmers can easily be several times as productive as the average, so their pay is well worth it, though of course only for those programmers. So my advice is to only outsource simple easily described stuff or alternately outsource an entire stand alone sub-system, that will require virtually no great interfacing effort. I personally do not agree with the approach where a large project is divided up over a huge programming army. For complex software "small teams are the only viable teams". It makes more sense to take a longer schedule with a small team than to think one can work faster with a huge team. Like scientific experiments, complex software cannot be hurried along. I remember helpless project managers always demanding more staff when projects started to run late. It was hopeless to put more staff on, as all it did was add to costs, but make no difference to schedule. A good example of the delivered benefits of small teams that I can quote, is the engineering team at the original Lockheed "skunk" works, where just 150 or so engineers designed and completely developed many of the world's best military aircraft in record time. "Small is beautiful"!

Hardware in an electronics environment is very dependent on being flexible and accepting that major changes are normal. In our early days electronics was based on designs where fairly large devices (eg. thermionic valves for microwave devices) had to be collected and installed on platforms where these individual devices would be connected to each other with fairly thick wires, as the devices needed larger currents. As they took up a large space, there were relatively few issues with cooling or mechanical strength. However fairly soon after that the transistor was invented, so the newer approach to design meant that the basic devices were collected onto a flat PCB (Printed Circuit Board), and the wiring in between devices was reduced to very thin tracks embedded inside the PCB. Connection between PCBs was through a chassis and back-plane onto which each PCB was plugged in using slide-in connectors. This system of PCBs, chassis and connectors has continued till today, but the PCBs have taken on enormous extra functionality as now many of the earlier circuits have been entirely replaced by small ICs (Integrated circuits) that can do a series of complete major functions within the IC itself. So to some extent system design has moved from the system companies to the semi-conductor IC designers.

So electronic circuit design at a systems company is becoming a much higher level activity. New issues with interference and very high frequencies can cause major failures as these devices now function at very high frequency but with very low voltages. When we were designing our spread spectrum radio, we had to be very careful about not interfering with the Radar and air traffic signals at nearby Dorval airport. These days power supplies and cooling are becoming more important in the design of the packaging as all this tightly packed stuff generates a large amount of heat. In some of our systems (for reliability reasons) we could not use cooling fans and had to rely on designs using metal mounts and innovative air flow schemes to keep the electronic equipment from over heating. So nowadays electronic designers also need good mechanical engineering skills to handle these issues. Even the base PCB card now has several layers of embedded track for the more complicated designs. So the PCB manufacturing facilities had to be kept on being upgraded with newer automatic component insertion machines and with automatic testers. All this means that electronic circuit designers have had to become more and more like systems programmers as they move to working at higher levels further away from the components. As I have said before, systems and electronic engineering is a constantly changing profession. Similarly the factory folk have to keep learning new methods too.

When it comes to the manufacture of electronic and other hardware, the approach that we followed at CAE and GEC (and I believe it is the same at other companies) was to organize around ease of manufacture. What this means is that all similar parts, from all the different projects, are lumped together into batches, so that each batch can be made by the same team. This keeps costs low and manufacture is fast . One consequence of this in a project company like CAE, is that we needed a well organized manufacturing planning system that reorganized the elements of different projects into these batches. When this is done the parts coming out of manufacture have to be tracked and separated out to the individual project test and assembly sites. The tracking and assembling document is called a family tree. It is just a breakdown of the system from the highest top level to greater and greater detailed levels. A special group of manufacturing planners are charged with the task of tracking progress and then ticking off parts as they are finished. Major purchased parts are also listed on the family tree but separately given to major purchase buyers who negotiate with external suppliers. Similarly the smaller assemblies of integrated circuits and components are ordered and kept in advance in stores, ready for call up when manufacturing. Hence an entire parallel paper process tracks the actual physical process to make sure nothing is forgotten or lost. It is quite a complicated process and somewhat difficult for a project person to find where any particular item on a project is in the schedule of process. So good project companies are a complex of networks of people interfacing with each other at various levels without paying much attention to the official hierarchy. Hence its very important in such companies to create these unofficial friendly networks of teams and people. That is the only way one can find out what is happening and then agitating for progress on your particular project parts.

Human nature being what it is, people are reluctant to change the well understood power structure of a traditional organisation. So systems companies try to modify their official organisations as little as possible, by keeping the old hierarchical arrangements. This leads to a hidden virtual organisation where everyone works around the official hierarchy by getting involved in the unofficial network just to get the work done properly. That is why it is so important for new employees to get to know how to work the "unofficial" network. It would be a great benefit if systems companies officially changed over to a network-like minimally layered organisation.

14.Thoughts on management for a technical project company

I think a great deal of success in management relies on common sense gained through experience. Most management issues end up being people related issues. There are also business related issues. Unfortunately this leads to the general view that the most important skill sets for managers relate to people, politics and business. This might be true in production or commodity product type companies, but it really is not suitable as a template for choosing managers and leaders in technical and project companies.

Even though people issues may be front and centre in technology companies, the really key issues are usually technical choices, and design approaches. If a manager is unable to guide or lead such efforts, he or she does not really contribute much other than some administrative effort. So all critical management positions should be staffed with the best engineers not the best "management types". This is true even at the highest levels. I have seen many examples of finance or business type leaders who have destroyed good technical companies because they did not understand the true value of what they might have inherited. In addition technical people do not really respect leadership provided by managers who can only deal with them at arms length. In the industrial world, ALL the entrepreneurs who created great companies were also superb technicians. The technology world is really no place for amateurs. Even sales departments have to use knowledgeable engineers not just salesmen. I have had potential clients , especially in the far east, complain to me that some (mostly American) companies sent only salesmen to talk to them, when they actually wanted an engineering expert to discuss their problems and issues.

Another myth that arises from time to time, is the one about good engineers being unable to also be good managers. I think it is just nonsense promulgated by "management types " trying to shore up their weak position. Even though it is only anecdotal, I have seen no correlation between lack of management skills and good technical skills. So on the basis of my own relatively few observations, I would always choose the best or maybe the most experienced engineer to be the leader and manager. As the saying goes someone should be able to "walk and chew gum at the same time. " Some professions are much more protective of their leadership requirements,eg lawyers and doctors, and keep control and management within their professions only. Again it is only reasonable that appropriate experience is what allows someone to guide and lead others.

Another important issue is the dedication to work that many technical people have. Technical work especially advanced technical work is by its nature quite fascinating and will absorb any serious engineer. So much so, I never had a problem of salary or work conditions with any of our engineering staff even though our company had very average benefits and rather mediocre salaries. As long as we did the most advanced technical projects our staff were content. As one of my colleagues once quipped " doing a project successfully just puts you in line for the chance to be allowed to work on the next project!" The only time in my life when I was totally unaware of the time was when I was deeply immersed in a technical problem, and could spend long hours without feeling at all tired. In later life I had to do a lot of business and marketing work as well as technical work. I would say that business and marketing is "exciting", but technical work is totally "absorbing".

Since I did a lot of work in Asia as well as America, I noticed an important difference in how Asian companies handle the issue of contracts. In Asia contracts are mostly based on trust in specific individuals, and they prefer to rely on a simple handshake. This made it crucial that the lead engineer who was going to design a project was someone in whom the client had complete confidence. Hence I always got heavily involved in marketing a system. In North America, and to some degree in South America, they had a much more formal legal structure for agreeing contracts. So during negotiations we had to go through a long list of contractual terms that had to be agreed to before we signed a contract. As this was normal business practice, the client expected it. In the East, if we had insisted on this approach, the clients would have been offended, so I had a terrible time persuading our internal contracts department to cut down the number of terms and conditions in the contracts. It is just another of the foolish handicaps of business practice in America. As a Malaysian client once said to me " If you don't perform, you will lose your reputation and our business for ever. If you do perform, we will return again and again. We want a working system, not some legalistic monetary recompense".

Actually come to think of it, this is true even in the west. One day we were chatting with our Irish client who had purchased a fossil power plant simulator from us.He told me that the reason why he had chosen us was because he completed trusted our lead engineer, even though we had never built a fossil power simulator before. Even some of our American customers told us in confidence later on, that they chose us over more experienced competitors, mostly because they liked and trusted us!! I would suggest that Business organizations cut back on the legal contractual bureaucracy, reducing terms and conditions to a simple single page of important terms and conditions. It would save a lot of money, and irritation. This type of behaviour does not even really protect one from much, as any sensible supplier company just has to perform, if it ever wants to do business in the future.

Another interesting management issue relates to the amount and method of Research and development (R&D) a project company should do. The organisation template of production type companies is not relevant. Production companies do the R&D and initial engineering using separate funds and staff to develop a prototype, test it and then release the drawings and instructions to the production departments to produce multiple copies. All the management and administration mythology is based around this template. This is not appropriate for project companies. Here engineering and development is best done as an initial part of the actual project. It is the only practical way, especially as most projects are fairly unique and different from each other. If there is a multiple production component, it is usually part of the project itself eg. building multiple ship sets for a particular navy ship type.   Conventionally the worth of R & D is measured just by how much money was spent, which is nonsense. It is the usual human tendency to put a value only on what is easy to count even if the measurement is not meaningful. That is why it is useful to have an actual real customer for the R & D, at least the customer will appreciate the work.

A corollary of this way of thinking is that there ought to be a radical rethink of formal education. So for instance it is debatable just how much direct use conventional engineering degrees are as the time in formal education usually lasts around 3 or 4 years after high school. There is very little use made of the actual course work in real world engineering. Most useful learning takes place on the job. It would be more useful if colleges gave a short summary type overview of the general sciences and mathematics. Then companies should allow staff to take short but precise courses whenever it is directly required at work. If we had such a distributed educational system, the formal course could be over in less than a couple of years and the short pertinent courses could each be about a few weeks or so in length. A major problem that occurs is that one soon forgets a great deal of early course work after some years, as I can easily testify. Many real world problems require self generated solutions where potted educations do not help much. I suspect we have the current long university education setup organised by professors to produce other professors rather than engineers. Working level engineering teams can successfully tackle the most complex problems, even when it is not clear why this is theoretically successful. It could be why science and theory often follow engineering. Instinct rather than scholarly theorizing often has better results. This may explain why most systems engineering and software advances occur in industry rather than in universities. Engineers do not always have to know why something works, just that it does.

A further advantage of this method of project R&D working is that any brand new research or development work has a definite paying customer, so it is never wasted. It is also focused and what the customer actually wants rather than a wish list generated at an arms length supplier. If the customers ' funds are not sufficient in a competitive situation, one can add ones' own company R&D funds to supplement the funds. This is a more sensible approach to R&D funding and direction than blue skies guess work. This implies a different actual management structure with R&D being part of the project and under the control of project engineers and project managers. Separate R&D management is unnecessary except perhaps to coordinate design issues across several projects. Some project companies have tried to do independent development of new technologies off from a real customer project. My personal observations are that in ALL the cases, that I have observed, it has mostly failed. Usually because one can rarely find an actual customer for the pre-invented technology. This highlights the advantage of working with early adapter customers. These customers have a good feel for which future technologies they will want to use.

So the management issue is to concentrate on finding such customers and following them closely, not wasting funds on company funded independent R&D. This is not to say that the work is not advanced and unique. It is just that it is completely relevant and cost effective. So flexibility becomes an important part of a project company's organisation, with key management and staff changing from project to project. This only requires a very thin layer of fairly static senior management, right at the top, to coordinate strategic items and finance. Ideally a project company needs very few layers of management, barely two or three. Any more, and an entrenched bureaucracy will form and gum things up.

In the last 15 years or so a new type of company operations management software system has become popular ie. manufacturing resource planning and control systems (MRP) , Enterprise Resource planning system (ERP) and the like. Essentially all the company operations and accounting data is kept in one centralised system and can be accessed throughout the company by different departments, of course subject to security rules. These systems have replaced separate independent computer systems each of which were developed at different times and so in different technologies over the years as departments computerised their operations. It was the same at CAE. The administrative systems that we had, did the job that was required and were reasonably efficient. However fashion dictated that CAE should have an ERP. Using the widely feared year 2000 possible crash as an excuse, the company decided to buy a brand new ERP for all the administrative functions of the company. It just so happened that I worked on that team as one of the staff that represented the company so I had a good look at what we obtained. I have to confess that I suspect it was mostly a waste of money though the final system was a lot more modern than the systems that were replaced. We got very little extra functionality for the expense and no cost savings whatsoever. Essentially such a system requires the company to spend a great deal of time studying complex operations and producing bureaucratic rules that are then enforced when data is put into the system using different types of display pages. After the data is input one can generate reams and reams of reports for management and operations. It is possible that such a system is a big benefit in a large production company, as bureaucratic rigidity will keep the large production volumes under better control. For a medium sized project company it just makes operations more expensive and rigid. It is probably better to encourage staff to just talk a lot more with each other, in order to keep things moving. Computer based bureaucratic systems will make quick changes much more difficult, in spite of the propaganda that it will be easier. This is mostly on account of the great deal of training that users have to be given and even then it is difficult to bypass the system in an emergency. The maintenance of such a complex system is also quite expensive. Since those days, I notice the ERP industry have dramatically cut their prices and made it possible to buy subsets of their systems rather than the whole company-wide system. In that case some parts of such a system might well be worthwhile.

A modern undesirable trend is to give excessive importance to financial issues in the running of a company. It is part of this destructive trend of financializing everything. I prefer to think about finance and accounting as support functions to business, certainly not their primary function. In my early days at work, that was exactly how organisations were run. In later years our newer executive management made several years of strategic mistakes by not paying proper attention to the real business issues related to actually making things. One of the consequences was that because of these foolish mistakes made by the newer management, CAE went heavily into debt. In a panic the company executives felt they had to sell off the control and power simulator businesses. At least they kept the "crown jewels" ie. flight simulators. It was typical short term thinking, as all 4 sold businesses had by then done all the required expensive technical and market developments for each one to be the acknowledged world leader in its own special field. So it was the buying companies that gained the benefit of CAE's efforts! I notice many years later CAE is once again trying to regain something like what it once had but it may be a bit too late.

The recipient companies are probably still leading their individual markets. Energy systems is the world leader in modern DMS systems. Power simulators are still the international leader in nuclear training simulation, as is the Marine control systems business in advanced naval machinery control. Even ATC systems are world leaders in the new technology of procedural oceanic ATC, ADS-B, satellite tracking and controller-pilot data link type ATC systems, which are on the way to replace the older radar technology based systems. In each of the above cases CAE overtook and replaced significantly larger companies, simply by taking the risk and implementing better and newer technology first. I have lost touch with what happened to my other old British company GEC (which is no more), but I hope the better technology sections are still thriving somewhere under new management.

A final thought. Management should not be afraid of mistakes and experiments in new ways of doing things. It is more or less the only way we humans innovate and how businesses grow, especially technical businesses. Book knowledge can only go so far. A vibrant company will have staff from all sorts of unusual disciplines and backgrounds. At CAE we were lucky to have a sympathetic management and a culturally international staff, so we grew organically at a very fast pace. In fact the company started to slow down in real innovative growth when we became more and more "normal". This means one has to continuously fight the bureaucratic tendency in all organisations to play safe. If this tendency wins, the organisation is bound to disappear or become irrelevant fairly soon. This issue is part of a phenomenon commonly called "a paradigm shift", for when something unusual happens then many of the successful old ways of working do not continue to give any advantage. When I joined CAE we were the smallest of the flight simulator suppliers, and virtually non existent in large and complex control systems. Ten years later we were the largest flight simulator supplier and a world leader in power simulators, energy and marine control systems. I believe the only reason was that we took risks, experimented and tried new approaches while our competitors stayed with the old conservative ways. The future is never safe and steady. As one can see from my own career, I have had scores of failures, but I learned something valuable from each failure. In the end we made lots of money from the few successes that resulted from what we learned by failing. So to modify the old NASA saying "FAILURE IS AN OPTION"!

Glossary.

ADA   A special programming language
AECL.....   Atomic Energy of Canada Ltd
AEDC....   Alexandria Electrcity Distribution Company
AGC......   Automatic Generation Control
Alpha.... . 64 bit computer from DEC company
APU......   Auxilliary Power Unit applies external power to aircraft.
ATC......   AirTraffic Control system
ATM......   Asynchronous Transfer Mode (Communication protocol)
Assembler Programming language closest to machine language
AYDIN   Company manufacturing electronic display systems

BAUD   Speed of communications in bits per second
Bomb   Jargon used for system crash
Bus   Electrical wired cable used to link electronic devices
BWR   Boiling Water Reactor, a type of nuclear reactor

C   A Standard programming language
CAD   Computer Aided Design
CANDU   Canadian designed nuclear reactor
Chassis   Box into which PCBs are collected and held
CISC   Complex Instruction Set Computer
Compiler   Program that converts between computer languages
CRT   Cathode Ray Tube
Core   A type of magnetic storage technology

DACBUS   Data Acquistion and Control Bus
Database Editor Program to set up a database (DBE)
DATAPATH   Trademarked name for a series of electronic interfaces
Data Logger   Computer system used to collect real time data
DCC   Digital Computer Control
Debug   Looking for errors in programs
DEC   Digital Equipment Company (a manufacturer of computers)
DECNET   Set of software programs to support communications
Display Generator Electronic chassis used to build graphic displays
DMC   Datapath Micro Computer
DMS   Distribution Management System (Controls power distribution
               system)
DP   Distribution Point (medium voltage substation)
Drum   Type of magnetic storage technology
DSP   Digital Signal Processor
DUSC   DACBUS-UNIBUS Smart Controller

Earthfault   Electrical short circuit to ground (a type of error)
ECM   Engine Control Module
ED   Economic Dispatch (specialised power allocation program)
EGAT   Electricity Generation Authority Thailand

EMS	Energy Management System (Controls power transmission system)
Ethernet	Type of wired link and protocol used to network computers
EEPROM	Electrically Eraseable PROM
ERP	Enterprise Resouce Planning system
Executive	Basic software that schedules programs
FAA	Federal Aviation Authority (USA)
FDDI	Fiber Distributed Digital Interface (Communication protocol)
Fortran	High level programming language
Front End	Special computer to manage communications
Full Graphics	Detailed Graphic technology providing great accuracy
Gas Turbine	Jet engine
GIS	Graphical Information System
GMS	Generation Management System (Controls power generators)
HMI	Human Machine Interface
HUMS	Health and Usage Monitoring System
HV	High Voltage
IBM	International Business Machines (manufactures computers)
IC	Integrated Chip
IDT	Integrated Display Technology (early supplier of full graphic systems)
IEC	Israel Electric Company
IEEE	Institute of Electrical and Electronic Engineering
IF	Instructor Facility
Interdata	Computer manufacturer
I/O	Input/Output
IPMS	Integrated Platform Management System
IRIG –B	Satellite clock signal
ISO	International Standards Organisation
JETS	Joint Enroute Terminal System (for ATC)
Kiosk	A small Electrical switching station
LAN	Local Area Network
LCD	Liquid Crystal Display
Master Station	Centralised control system
MMI	Man Machine Interface
MW	Mega Watt
Microwave	Technology used to transmit Radar / wireless signals
Mil Spec	Military Specifications (used for defining military devices)
Mimic	Large Panel that uses lights and buttons to show layout of devices
Modeller	Engineer who develops software that matches a physical system
Multiplexer	Technology to share electronic signals through a single device

NARI	Nanjing Automation Research Institute
NASA	National Aeronautic and Space Administration (USA)
Network	Collection of computers and electronic chassis connected electrically.

OPEN	Computer system Standards that anyone can use freely
Operating System	Core Sofware to manage a computer
ORACLE	A standard relational database system
Oscilloscope	Instrument to measure electronic signals

PC	Personal Computer
PCB	Printed Circuit Board
PDP 11	A 16 bit computer made by DEC
PLC	Programmable Logic Controller
PLM	A programming Language
Point	Analog or digital signal being measured or calculated
POSIX	Computer standard for operating systems
Processor	Another name for computer type device
PROM	Programmable Read Only Memory
PWR	Pressurized Water Reactor (type of nuclear reactor)

QA	Quality Assurance

R and M	Reliability and Maintainability
RADAR	Technology used to detect remote moving objects
RAM	Type of memory
RAMTEK	Display system manufacturer
Raster Graphics	Technology using line scanning to draw objects on CRT screen
RDBMS	Relational Database Management System
RFP	Request For Proposal
RISC	Reduced Instruction Set Computer
RMX	Computer operating system software
ROM	Type of memory
RSX	Computer operating system software
RS 485/ 422	An electrical wiring standard for copper LANs.
RTU	Remote Terminal Unit, Electronics used to monitor/control devices at a distance from control center

SCADA	Supervisory Control And Data Acqusition (system)
SER	Sequence Of Events Recorder
SHINMACS	Ship Integrated Machinery Control System
SHINPADS	Ship Integrated Processing And Display System
Silicon Graphics	Computer and display system manufacturer
Snag	Jargon for computer software error
SYBASE	A standard relational database system

TB	Terminal Blocks (used for connecting wires between devices)
TI	Texas Instruments (computer manufacturer)

Telecontrol	Technology for remote control
Telemetry	Technology used to measure and control remotely
Thermodynamics	Science of heat and Energy
Token Ring	Technology used for networking
Track Circuit	Railway technology used to signal track occupancy
Transducer	Device to convert a physical change into an electrical change.

UNIX	A standard operating system
UPS	Uninterruptible Power Supply

VAX	32 bit computer from DEC
Vector graphics	Display graphics by direct point to point drawing
VDU	Visual Display Unit
VME	Motorola range of electronic circuits.
VMS	Virtual Memory System (operating system for VAX computers)

WDT	Watch Dog Timer
WINTEL	Acronym for Windows/Intel type PC computers

X-Ref	Acronym for Cross Reference (a shared memory database)

YVPO	Yangtse Valley Plannng Office

www.ingramcontent.com/pod-product-compliance
Lightning Source LLC
Chambersburg PA
CBHW051528170526
45165CB00002B/645